Collective Volumes, Cumulative Indices, and Annual Volumes 64–68 are available from John Wiley & Sons, Inc.

ORGANIC SYNTHESES

ORGANIC SYNTHESES

AN ANNUAL PUBLICATION OF SATISFACTORY
METHODS FOR THE PREPARATION
OF ORGANIC CHEMICALS

VOLUME 68

1990

BOARD OF EDITORS

JAMES D. WHITE, *Editor-in-Chief*

JOHN WILEY & SONS

NEW YORK · CHICHESTER · BRISBANE · TORONTO · SINGAPORE

547
068

Published by John Wiley & Sons, Inc.

Copyright © 1990 by Organic Syntheses, Inc.

All rights reserved. Published simultaneously in Canada.

Library of Congress Catalog Card Number: 21-17747
ISBN 0-471-53789-6

Printed in the United States of America

10 9 8 7 6 5 4 3 2 1

243242

NOTICE

With Volume 62, the Editors of *Organic Syntheses* began a new presentation and distribution policy to shorten the time between submission and appearance of an accepted procedure. The soft cover edition of this volume is produced by a rapid and inexpensive process, and is sent at no charge to members of the Organic Divisions of the American and French Chemical Society, The Perkin Division of the Royal Society of Chemistry, and The Society of Synthetic Organic Chemistry, Japan. The soft cover edition is intended as the personal copy of the owner and is not for library use. A hard cover edition is published by John Wiley and Sons Inc. in the traditional format, and differs in content primarily in the inclusion of an index. The hard cover edition is intended primarily for library collections and is available for purchase through the publisher. Annual Volumes 60–64 have been incorporated into a new five-year version of the collective volumes of *Organic Syntheses* which appears as *Collective Volume Seven* in the traditional hard cover format. It is available for purchase from the publishers. The Editors hope that the new *Collective Volume* series, appearing twice as frequently as the previous decennial volumes, will provide a permanent and timely edition of the procedures for personal and institutional libraries. The Editors welcome comments and suggestions from users concerning the new editions.

NOMENCLATURE

Both common and systematic names of compounds are used throughout this volume, depending on which the Editor-in-Chief felt was more appropriate. The *Chemical Abstracts* indexing name for each title compound, if it differs from the title name, is given as a subtitle. Systematic *Chemical Abstracts* nomenclature, used in both the 9th and 10th Collective Indexes for the title compound and a selection of other compounds mentioned in the procedure, is provided in an appendix at the end of each preparation. Registry numbers, which are useful in computer searching and identification, are also provided in these appendixes. Whenever two names are concurrently in use and one name is the correct *Chemical Abstracts* name, that name is preferred.

SUBMISSION OF PREPARATIONS

Organic Syntheses welcomes and encourages submission of experimental procedures which lead to compounds of wide interest or which illustrate important new developments in methodology. The Editorial Board will consider proposals in outline format as shown below, and will request full experimental details for those proposals which are of sufficient interest. Submissions which are longer than three steps from commercial sources or from existing *Organic Syntheses* procedures will be accepted only in unusual circumstances.

Organic Syntheses Proposal Format

1) Authors
2) Literature reference or enclose preprint if available.
3) Proposed sequence
4) Best current alternative(s)
5) a. Proposed scale, final product:
   b. Overall yield:
   c. Method of isolation and purification:
   d. Purity of product (%):
   e. How determined?

6) Any unusual apparatus or experimental technique:
7) Any hazards?
8) Source of starting material?
9) Utility of method or usefulness of product.

Submit to: Dr. Jeremiah P. Freeman, Secretary
Department of Chemistry
University of Notre Dame
Notre Dame, IN 46556

Proposals will be evaluated in outline form, again after submission of full experimental details and discussion, and, finally by checking experimental procedures. A form that details the preparation of a complete procedure (Notice to Submitters) may be obtained from the Secretary.

Additions, corrections, and improvements to the preparations previously published are welcomed; these should be directed to the Secretary. However, checking of such improvements will only be undertaken when new methodology is involved. Substantially improved procedures have been included in the Collective Volumes in place of a previously published procedure.

ACKNOWLEDGMENT

Organic Syntheses wishes to acknowledge the contributions of E. I. du Pont de Nemours and Co., Inc., Hoffmann-La Roche, Inc., and the Rohm and Haas Co. to the success of this enterprise through their support, in the form of time and expenses, of members of the Boards of Directors and Editors.

PREFACE

Annual volume 68 contains thirty one checked and edited procedures representative of some of the most active areas at the frontier of organic synthesis. Although there is no single theme to this volume, the preparations are broadly compiled into six categories, viz (a) simple one- and two-carbon synthons, (b) chiral auxiliaries and stereocontrolled processes, (c) functionalizations employing organometallic reagents, (d) useful starting materials for constructing complex molecules, (e) emerging reactions of potentially broad utility and (f) cryptands with novel host-guest properties.

The first six contributions are directed toward small but versatile structures that have general utility in synthesis. **TRIMETHYLSILYLDIA-ZOMETHANE,** prepared by the reaction of trimethylsilylmagnesium chloride with diphenyl phosphorazidate, is a safe, convenient substitute for the hazardous diazomethane, and **(PHENYLTHIO) NITROMETH-ANE,** which illustrates both the synthesis of phenylsulfenyl chloride and its reaction with the anion of nitromethane, is an important reagent for the construction of heterocycles. A carefully documented procedure describes **LITHIUM ACETYLIDE** by a route that avoids the troublesome dilithio species, and the preparation of **ACETYLTRIMETHYLSILANE** via lithiation of methyl vinyl ether exemplifies a general entry to the important class of acylsilanes. The last two contributions in this section demonstrate cycloadditions of dichloroketene, in one case to an alkyne to yield, after reductive dechlorination, **3-BUTYLCYCLOBUTEN-ONE,** and in the other to a cycloalkene to afford, after oxidative cleavage of the derived cyclobutanone, **cis-1-METHYLCYCLOPENTANE-1,2-DICARBOXYLIC ACID.**

A second set of six procedures features compounds of value for their stereochemical properties. **(S)-(−)-METHYL p-TOLYL SULFOXIDE** of high enantiomeric purity is obtained by asymmetric oxidation of the corresponding sulfide, and this preparation is followed by two stereoselective reductions. In the first, reduction of the meso structure, 2,2-dimethylcyclohexane-1,3-dione, is carried out with baker's yeast to give **(S)-(+)-3-HYDROXY-2,2-DIMETHYLCYCLOHEXANONE** in what is likely to be the forerunner of a genre of microbial reactions, and in the second a directed, homogeneous hydrogenation of an allylic alcohol is accomplished with a bisphosphinorhodium catalyst to produce **METHYL anti-3-HYDROXY-2-METHYLPENTANOATE.** Three additional preparations in this group describe the widely used chiral auxiliary **(S)-4-(PHENYLMETHYL)-2-OXAZOLIDONE,** together with its application in an asymmetric aldol condensation to furnish **(2S,3S)–3-HYDROXY-3-PHENYL-2-METHYLPROPANOIC ACID,** and **1,4-DI-O-BENZYL-L-THREITOL,** an auxiliary obtained from tartaric acid that has been used to stereochemically direct reactions such as cyclopropanation.

The introduction and transformation of functional groups through the intermediacy of organometallic species has come to play a vital role in synthesis, and this area is represented by seven procedures, five of which make use of either palladium or tin (or, in one case, both). The first preparation of this series is that of **ALLYLTRIBUTYLTIN,** a widely used reagent in transmetallation reactions and in substitutions that transfer an allyl group from tin to carbon. This is followed by a palladium(II) mediated allylic acetoxylation leading to **2-CYCLOHEPTEN-1-YL ACETATE** and by syntheses of **4-tert-BUTYL-1-VINYLCYCLOHEX-ENE** and **1-(4-tert-BUTYLCYCLOHEXEN-1-YL)-2-PROPEN-1-ONE** that employ coupling of a vinyl triflate with a vinylstannane in the presence of palladium(0). In a similar vein, the palladium(0) promoted coupling of a vinylboronate with a vinyl bromide is demonstrated in the preparation of **(1Z,3E)-1-PHENYL-1,3-OCTADIENE,** and reduction of a vinyl triflate, via a σ-vinylpalladium complex, is detailed in the synthesis of **CHOLESTA-3,5-DIENE** from cholest-4-en-3-one. A mercurative sulfonylation is the pivotal step in the synthesis of **(2-PHENYLSUL-FONYL)-1,3-CYCLOHEXADIENE,** a valuable Diels-Alder and Michael reaction partner and, in a high-yielding transesterification protocol that is especially useful for the acquisition of esters of sterically hindered alcohols, the lithium alkoxide from menthol is used to obtain **(−)-MENTHYL CINNAMATE** and **(−)-MENTHYL NICOTIN-ATE.**

Organic Syntheses has traditionally provided preparations of common starting materials for multistep approaches to complex structures and, although the importance of this aspect of the series has somewhat diminished with the advent of more extensive commercial catalogs, there remains a need for reliable routes to certain compounds that cannot be purchased or are prohibitively expensive. **3-METHYL-2(5H)-FURA-NONE,** a useful butenolide accessible from furfural via 2(5H)-furanone, is one such material, and **BICYCLO[3.3.0]OCTANE-2,6-DIONE,** derived from transannular cyclization of cycloocta-1,5-diene with iodosobenzene diacetate, is another. The four-carbon vinylsilane **(Z)-4-(TRIMETHYLSILYL)-3-BUTEN-1-OL** also belongs to this class, and its application to heterocycle synthesis is exemplified in the companion preparation of **1-(4-METHOXYPHENYL)-1,2,5,6-TETRAHYDRO-PYRIDINE.** A more elaborate starting material **3,3a,3b,4,6a,7a-HEXAHYDRO-3,4,7-METHENO-7H-CYCLOPENTA[a] PENTA-LENE-7,8-DICARBOXYLIC ACID** that has been used for building a variety of cage structures, including dodecahedrane, is expressed as the "Domino Diels-Alder Reaction" product of 9,10-dihydrofulvalene.

Although this volume does not emphasize specific reactions, three processes that deserve special mention are included. These are an immonium ion-based Diels-Alder reaction that is carried out in an aqueous medium and which leads to **N-BENZYL-2-AZANORBORNENE,** the seldom-used but versatile Carroll rearrangement of an allylic β-keto ester to give **5-DODECEN-2-ONE,** and the Skattebøl rearrangement of a 1,1-dibromocyclopropane to provide **(1R)-9,9-DIMETHYLTRICY-CLO[6.1.1.0^{2,6}]DECA-2,5-DIENE,** a chiral homolog of isodicyclopentadiene.

The volume concludes with four procedures describing two different classes of cryptands. The first, **4,13-DIAZA-18-CROWN-6,** is the prototype of an interesting family of "two-armed macrocycles," and the last three entries represent a series of homologous calixarenes each derived from the condensation of formaldehyde with p-tert-butylphenol. The syntheses of **p-tert-BUTYLCALIX[4]ARENE, p-tert-BUTYLCA-LIX[6]ARENE,** and **p-tert-BUTYLCALIX[8]ARENE** have been worked out so that any one of these substances can be obtained free of homologs with minimal purification.

The procedures published in *Organic Syntheses* are selected not only for their value in providing specific information to practitioners of synthesis but, in addition, they are intended to exemplify the highest experimental standards. These standards are ensured by the careful checking process carried out by the Board of Editors and their students, and it is hoped that the synthetic community at large will share my appreciation of those whose time and effort was consumed by the rigorous protocol involved in checking procedures for this volume. Their reward is the satisfaction of seeing an *Organic Syntheses* procedure used and reused in a way that advertises its accuracy and reliability.

As always, submitted procedures for inclusion in future volumes of *Organic Syntheses* are welcomed (a sample proposal at the end of this volume is intended to serve as a guide). In addition, the Board of Editors will continue to solicit appropriate procedures suggested by the readership. The criteria for selection of a particular preparation can be inferred from the contents of this and preceding volumes of *Organic Syntheses*, but it is important to emphasize that, even where the principal purpose is to illustrate a general procedure, its application to a specific compound should be described.

Finally, I note that the production of this, as with previous volumes of the series, has been greatly assisted by the steadfast contributions of the Secretary to the Board, Professor Jeremiah P. Freeman, and our Assistant Editor, Dr. Theodora W. Greene. Their expertise and painstaking efforts with textual material and graphics is largely responsible for transforming an assorted collection of manuscripts into the volume before you.

JAMES D. WHITE

Corvallis, Oregon
November, 1988

ORGANIC SYNTHESES

TRIMETHYLSILYLDIAZOMETHANE

(Silane, (diazomethyl)trimethyl-)

$$(CH_3)_3SiCH_2Cl \quad \xrightarrow{Mg} \quad (CH_3)_3SiCH_2MgCl$$

$$(CH_3)_3SiCH_2MgCl \quad \xrightarrow[\text{2) } H_2O]{\text{1) } (PhO)_2P(O)N_3} \quad (CH_3)_3SiCHN_2 \; + \; (PhO)_2P(O)NH_2$$

Submitted by Takayuki Shioiri, Toyohiko Aoyama, and Shigehiro Mori.[1]

Checked by George Maynard and Leo A. Paquette.

1. Procedure

A. *Trimethylsilylmethylmagnesium chloride.* Magnesium turnings (10.7 g, 0.44 g-atom) are placed in a dry, 300-mL, four-necked, round-bottomed flask equipped with a Teflon-coated magnetic stirring bar, 200-mL pressure-equalizing dropping funnel capped with a rubber septum, thermometer, rubber septum, and a reflux condenser connected to an argon flow line. The apparatus is flushed with argon, and an argon atmosphere is maintained throughout the reaction. Anhydrous diethyl ether (40 mL) (Note 1) and 0.1 mL of 1,2-dibromoethane are placed in the reaction flask with a syringe, and the mixture is stirred at room temperature for 15 min. The dropping funnel is charged with a solution of 45.4 g (0.37 mol) of chloromethyltrimethylsilane (Note 2) in 100 mL of anhydrous diethyl ether with a syringe. With stirring, about 10 mL of this solution is added at once. When the reaction has started (Note 3), the remaining solution is added dropwise at such a rate that a gentle reflux

1

is maintained throughout the addition (addition time about 2 hr). After the exothermic reaction subsides, the resulting solution is stirred at reflux by heating for an additional 1 hr. The reaction mixture is cooled to room temperature, and is used in step B.

B. *Trimethylsilyldiazomethane* (Note 4). A dry, 1-L, four-necked, round-bottomed flask is equipped with a liquid paraffin-sealed mechanical stirrer (Note 5), rubber septum, 200-mL pressure-equalizing dropping funnel capped with a rubber septum, and a reflux condenser connected to an argon flow line. The apparatus is flushed with argon, and an argon atmosphere is maintained throughout the reaction. A solution of 91.2 g (0.33 mol) of diphenyl phosphorazidate (Note 6) in 350 mL of anhydrous diethyl ether is placed in the flask with a syringe. The rubber septum is quickly replaced by a low-temperature thermometer. The Grignard reagent prepared in step A is transferred to the dropping funnel with a syringe. The flask is cooled with an ice-sodium chloride bath, and the stirring is started. When the internal temperature reaches -10°C, the Grignard reagent is added dropwise at such a rate that the internal temperature is maintained below 0°C (addition time about 1.5 hr) (Note 7). After the addition is complete, the ice-salt bath is replaced with an ice bath and the mixture is stirred for 2 hr, then allowed to stand in the ice bath for 14-16 hr. The reaction mixture is cooled again to -15°C with an ice-sodium chloride bath. With vigorous stirring, 35 mL of cold water is carefully added dropwise at such a rate that the internal temperature is maintained below 0°C (addition time about 1 hr), and the stirring is continued for 0.5 hr (Note 8). The reaction mixture is then filtered by suction using a glass filter. The white solid is thoroughly washed with three 100-mL portions of diethyl ether. The combined filtrate is washed with two 100-mL portions of cold water and dried over anhydrous sodium sulfate. After the sodium sulfate is removed by filtration, the filtrate is placed in a 1-L, round-bottomed flask equipped with a Teflon coated magnetic stirring bar and a 30-cm Vigreux column (15-mm diameter). With constant stirring by a magnetic stirrer, the solution is slowly concentrated to a volume of about 200 mL by distillation at atmospheric pressure with the bath temperature below 45°C. The concentration time requires about 6 hr (Notes 9 and 10). The remaining deep yellow solution is distilled through the same apparatus under reduced pressure between 100 mm at 0°C (bath temperature) and 15 mm at 40°C (bath temperature) until no more volatile material comes over, and the distillate is collected in a receiver cooled in a dry ice-acetone bath. The distillate is dried again over anhydrous sodium sulfate. The drying agent is removed by filtration, and 100 mL of hexane (Note 11) is added to the filtrate. In a manner similar to that described above, this solution is slowly concentrated by distillation at atmospheric pressure through a 30-cm Vigreux column (15-mm diameter). The color of the distillate gradually becomes yellow. Concentration is continued until the temperature of the vapor reaches 68°C (final oil bath temperature 87°C). The concentration requires about 3 hr (Note 9). Approximately 80-110 mL of the remaining yellow hexane solution contains 220-230 mmol of trimethylsilyldiazomethane (67-70% yield based on diphenyl phosphorazidate) (Notes 12, 13). This hexane solution can be stored without decomposition for periods exceeding 6 months at 0°C with protection from light.

2. Notes

1. Diethyl ether was distilled from lithium aluminum hydride under argon before use.

3

2. Chloromethyltrimethylsilane, obtained from Petrarch Systems Inc., was purified by distillation at 97°C under atmospheric pressure.

3. If the reaction does not start, the flask is gently heated by a heat gun.

4. *Trimethylsilyldiazomethane is both non-explosive and non-mutagenic.*[2] *Therefore, the very careful operations*[3] *used for the preparation of diazomethane are not necessary.*

5. A glass or Teflon stirring rod should be used.

6. Diphenyl phosphorazidate, obtained from either Daiichi Pure Chemicals Co., Ltd. or Aldrich Chemical Company, Inc., was purified by distillation under reduced pressure; bp 134-136°C/0.2 mm. It can be easily prepared according to *Org. Synth.* **1984**, *62*, 187.

7. After approximately two-thirds of the Grignard reagent is added, a large amount of white precipitate appears.

8. The mixture becomes yellow which is the color of trimethylsilyldiazomethane.

9. The color of the distillate is pale yellow because of co-distillation of trimethylsilyldiazomethane. Therefore, the rate of concentration is very important. If the rate of concentration is more rapid, the yield of trimethylsilyldiazomethane will decrease.

10. The submitters report that when a 30-cm Widmer column is used, the concentration time required is shorter (about 4 hr).

11. Hexane was purified by distillation.

12. The infrared spectrum (hexane) has absorptions at 2075, 1260, and 855 cm^{-1}. The ^{1}H NMR spectrum is as follows (hexane) δ: 0.16 (s, 9 H), 2.58 (s, 1 H) (internal chloroform standard).

13. The ^{1}H NMR of the hexane solution showed the presence of a trace of diethyl ether. The concentration of trimethylsilyldiazomethane was determined by the ^{1}H NMR analysis as follows: Ninety-one milligrams of dibenzyl was dissolved in 1 mL of a hexane solution of trimethylsilyldiazomethane, and its ^{1}H NMR spectrum was determined. The concentration (x mmol/mL) of trimethyl-silyldiazomethane was calculated as follows:

$$x = 2b/a \ (\text{mmol/mL})$$

a = Integral value (mm) of the methylene protons

$(\delta: \ 2.99)$ of dibenzyl.

b = Integral value (mm) of the methine proton

$(\delta: \ 2.58)$ of trimethylsilyldiazomethane.

3. Discussion

Trimethylsilyldiazomethane, as a stable and safe substitute for hazardous diazomethane, is useful both as a reagent for introducing a C_1-unit and as a C-N-N synthon for the preparation of azoles.[4] Many methods are described in the literature for the preparation of trimethylsilyldiazomethane, including the trimethylsilylation of diazomethane (7-74%),[5] the alkaline decomposition of N-nitroso-N-(trimethylsilylmethyl)amides (25-61%)[2,6] and the diazo group transfer reaction of trimethylsilylmethyllithium with p-toluenesulfonyl azide (38%).[7] The present modified diazo group transfer method appears to be the most practical, high-yield, and large scale procedure for the preparation of trimethylsilyldiazomethane.[8]

Diphenyl phosphorazidate can be replaced with diethyl phosphorazidate in the above procedure. Use of other azides such as p-toluenesulfonyl azide, p-methoxybenzyloxycarbonyl azide, diphenylphosphinic azide, or diphenylthio-phosphinic azide is less satisfactory. No reaction occurs when trimethylsilyl azide, dimethylthiophosphinic azide, or alkaline azides are used, while decomposition of formed trimethylsilyldiazomethane seems to occur when methanesulfonyl azide is used.[9]

The present procedure affords a general method for preparing silyldiazo-methanes from the corresponding chloromethylsilyl compounds. Typical examples are as follows:

$$CH_3$$
$$Ph-Si-CHN_2 \quad 76\%[9]$$
$$CH_3$$

$$Oi\text{-}Pr$$
$$CH_3-Si-CHN_2 \quad 62\%[9]$$
$$Oi\text{-}Pr$$

$$CH_3 \quad CH_3$$
$$CH_3-Si——Si-CHN_2 \quad 50\%[10]$$
$$CH_3 \quad CH_3$$

1. Faculty of Pharmaceutical Sciences, Nagoya City University, Tanabe-dori, Mizuho-ku, Nagoya 467, Japan.

2. Aoyama, T.; Shioiri, T. *Chem. Pharm. Bull.* **1981**, *29*, 3249.

3. Moore, J. A.; Reed, D. E. *Org. Synth., Collect. Vol. 5* **1973**, 352.

4. Shioiri, T.; Aoyama, T. *Yuki Gosei Kagaku Kyokaishi* **1986**, *44*, 149; *Chem. Abstr.* **1986**, *104*, 168525q.

5. Lappert, M. F.; Lorberth, J.; Poland, J. S. *J. Chem. Soc. (A)* **1970**, 2954; Martin, M. *Synth. Commun.* **1983**, *13*, 809.

6. Seyferth, D.; Menzel, H.; Dow, A. W.; Flood, T. C. *J. Organomet. Chem.* **1972**, *44*, 279; Crossman, J. M.; Haszeldine, R. N.; Tipping, A. E. *J. Chem. Soc., Dalton Trans.* **1973**, 483; Sheludyakov, V. D.; Khatuntsev, G. D.; Mironov, V. F. *Zh. Obshch. Khim.* **1969**, *39*, 2785; *Chem. Abstr.* **1970**, *72*, 111553p; Schöllkopf, U.; Scholz, H.-U. *Synthesis* **1976**, 271.

7. Barton, T. J.; Hoekman, S. K. *Synth. React. Inorg. Met.-Org. Chem.* **1979**, *9*, 297.

8. Mori, S.; Sakai, I.; Aoyama, T.; Shioiri, T. *Chem. Pharm. Bull.* **1982**, *30*, 3380.

9. Unpublished observations, these laboratories.

10. Sekiguchi, A.; Ando, W. *Chem. Lett.* **1983**, 871.

11. Ando, W.; Tanikawa, H.; Sekiguchi, A. *Tetrahedron Lett.* **1983**, *24*, 4245.

Appendix

Chemical Abstracts Nomenclature (Collective Index Number);
(Registry Number)

Chloromethyltrimethylsilane: Silane, (chloromethyl)trimethyl- (8, 9); (2344-80-1)

Diphenyl phosphorazidate: Phosphorazidic acid, diphenyl ester (8, 9); (26386-88-9)

Trimethylsilyldiazomethane: Silane, (diazomethyl)trimethyl- (8, 9); (18107-18-1)

(PHENYLTHIO)NITROMETHANE

(Benzene, [(nitromethyl)thio]-)

A. $\quad$ PhSH $\quad\xrightarrow[\text{Et}_3\text{N}]{\text{SO}_2\text{Cl}_2}\quad$ PhSCl

B. $\quad$ PhSCl $\quad\xrightarrow[\text{EtOH}]{\text{NaCH}_2\text{NO}_2}\quad$ PhSCH$_2$NO$_2$

Submitted by Anthony G. M. Barrett, Dashyant Dhanak, Gregory G. Graboski, and Sven J. Taylor.[1]

Checked by Zhou Zu-liang and Ekkehard Winterfeldt.

1. Procedure

CAUTION: Thiophenol (stench!!) and sulfuryl chloride are highly toxic. Steps A and B should be carried out in an efficient fume hood while wearing gloves and adequate eye protection. Sodium nitromethylate is explosive when dry and should be handled only as a slurry.

A. *Phenylsulfenyl chloride.* A 250-mL, three-necked, round-bottomed flask, fitted with a nitrogen inlet, pressure-equalizing 125-mL dropping funnel, and magnetic stirring bar, is charged with thiophenol (21 mL) (Note 1), dry triethylamine (0.25 mL) and dry pentane (100 mL) (Note 2) under a blanket of nitrogen. The remaining neck of the flask is stoppered and the nitrogen is allowed to sweep gently through the flask and out of the pressure-equalizing dropping funnel. The flask and its contents are cooled to 0°C with an ice bath and stirring is begun. The dropping funnel is charged with sulfuryl chloride (19 mL) (Note 1). The sulfuryl chloride is added dropwise

8

over a period of 1 hr to the chilled thiophenol solution with stirring. During this addition, a thick layer of white solid forms. It gradually dissolves as it is broken down. After the addition is complete, the ice bath is removed and the mixture is allowed to stir for 1 hr longer while slowly warming to room temperature. During the course of the addition and subsequent stirring, the clear, pale yellow solution becomes dark orange-red. The dropping funnel is replaced with an outlet adapter connected to a vacuum pump and the nitrogen inlet is exchanged for a ground glass stopper. The pentane and excess sulfuryl chloride are removed under reduced pressure at room temperature. The outlet adapter is replaced by a short-path distillation apparatus adapted for use under reduced pressure. The oily red residue is distilled to give phenylsulfenyl chloride as a blood-red liquid (26 g, 87%), bp 41-42°C (1.5 mm) (Note 3). This compound is stored under nitrogen until used in part B (Note 4).

B. *(Phenylthio)nitromethane.* Freshly cut sodium metal (4.8 g) is added to absolute ethanol (100 mL) in a 500-mL Erlenmeyer flask with a ground glass joint and allowed to react until the metal is completely consumed (Note 5). To this mixture is added a solution of nitromethane (12 g) (Note 6) in absolute ethanol (100 mL) with swirling. The phenylsulfenyl chloride (prepared earlier) is quickly poured into a 1000-mL, three-necked, round-bottomed flask fitted with a mechanical stirrer, nitrogen inlet/outlet adapter and a calcium chloride drying tube, and diluted with dry tetrahydrofuran (THF) (250 mL) (Note 7). Stirring is begun, the drying tube is removed, and the sodium nitromethane-ethanol slurry is added quickly in one portion to the THF solution (Note 8). The deep red solution immediately turns yellow and stirring is continued for a further 10 min. The reaction mixture consists of solid and liquid. It is dissolved in 200 mL of a 1 N sodium hydroxide solution and poured into a 1000-mL separatory funnel. Dichloromethane (200 mL) is added (Note 9), the aqueous layer (lower layer) is separated and the organic layer further extracted with 1 N sodium hydroxide (2 x 100 mL) (Note 10). The combined aqueous layers are washed with dichloromethane (500 mL) and acidified to pH 3 using 1 N hydrochloric acid. The brown organic layer that appears is separated, diluted with 50 mL of dichloromethane, dried over magnesium sulfate, filtered and concentrated at water aspirator pressure to give 18-19 g (60-65%) of crude (phenylthio)nitromethane as an orange-red oil. This material is of sufficient purity for many purposes. Further purification may be effected by distilling at reduced pressure to give (phenylthio)nitromethane (14 g, 50%) as a pale yellow oil, bp 85-95°C/0.05 mm (Notes 11-14).

2. Notes

1. Thiophenol (97%) and sulfuryl chloride (97%) were obtained from the Aldrich Chemical Company, Inc. and used without further purification.

2. Both pentane and triethylamine were obtained from the Aldrich Chemical Company, Inc. Before use they were dried over sodium wire and distilled from fresh sodium wire onto 4 Å molecular sieves under an atmosphere of nitrogen.

3. The submitters report isolated yields of phenylsulfenyl chloride of 82-92%.

4. Phenylsulfenyl chloride decomposes rapidly in moist air, and should be handled and stored under dry nitrogen.

5. The reaction is exothermic; cooling in an ice bath may be necessary to prevent the ethanol from refluxing.

10

6. Nitromethane was obtained from the Eastman Kodak Chemical Company and used without further purification.

7. Tetrahydrofuran was obtained from the Baker Chemical Company and distilled from sodium benzophenone ketyl before use.

8. The submitters preferred to transfer the phenylsulfenyl chloride via cannula. The sodium nitromethane-ethanol slurry is added by attaching the Erlenmeyer flask containing it to the reaction flask via an angle adapter and then simply inverting the Erlenmeyer flask. In either method additional absolute ethanol may be necessary to complete the latter addition.

9. Dichloromethane was purchased from the Baker Chemical Company and used without further purification.

10. The checkers used sodium hydroxide (3 x 100 mL) in one run which improved the yield about 5%.

11. During the distillation gases are evolved.

12. A pump of sufficient capacity must be used to maintain reduced pressure of at least 0.10 mm or extensive decomposition results.

13. (Phenylthio)nitromethane has the following properties: ^{1}H NMR (CDCl$_3$, 90 MHz) δ: 5.45 (s, 2 H, CH$_2$), 7.25-7.5 (m, 5 H, aromatic); IR (film) cm^{-1}: 3060 m, 3025 m, 2960 m, 2905 m, 1960 w, 1885 w, 1810 w, 1550 s, (NO$_2$), 1485 s, 1475 s, 1440 s, 1390 s, 1355 s, (NO$_2$), 1260 s, 1185 s, 1070 m, 1025 m, 1005 w, 900 m, 805 m, 745 s, 690 s.

14. *Storage and handling of (phenylthio)nitromethane.* Although (phenylthio)nitromethane will slowly decompose at room temperature, the submitters have found that the compound may be stored essentially unchanged in a freezer at -25°C. Since it has an unpleasant odor, it is best handled in a well ventilated hood; any spillage may be cleaned up with commercial bleach.

3. Discussion

(Phenylthio)nitromethane is a convenient reagent for the synthesis of derivatives of 3-methylfuran,[2] for the preparation of α-substituted phenylthio esters via the homologation of aldehydes,[3] and for the preparation of bicyclic β-lactams from monocyclic precursors.[4] This method is an adaption of Seebach's procedure.[5] Alternatively, (phenylthio)nitromethane may be prepared from the nitration of the dianion derived from (phenylthio)acetic acid[2] or from ethyl nitroacetate and N-(phenylthio)morpholine.[6] Neither of these procedures are as convenient on a large scale.

1. Department of Chemistry, Northwestern University, Evanston, IL 60201.

2. Miyashita, M.; Kumazawa, T.; Yoshikoshi, A. *J. Org. Chem.* **1980**, *45*, 2945.

3. Barrett, A. G. M.; Graboski, G. G.; Russell, M. A. *J. Org. Chem.* **1986**, *51*, 1012.

4. Barrett, A. G. M.; Graboski, G. G.; Russell, M. A. *J. Org. Chem.* **1985**, *50*, 2603.

5. Seebach, D.; Lehr, F. *Helv. Chim. Acta* **1979**, *62*, 2239.

6. Bordwell, F. G.; Bartmess, J. E. *J. Org. Chem.* **1978**, *43*, 3101.

Appendix

Chemical Abstracts Nomenclature (Collective Index Number);
(Registry Number)

(Phenylthio)nitromethane: Benzene, [(nitromethyl)thio]- (9); (60595-16-6)

Thiophenol: Benzenethiol (8,9); (108-98-5)

Sulfuryl chloride (8,9); (7791-25-5)

Sodium nitromethylate: Methane, nitro-, ion(1-), sodium (8,9); (25854-38-0)

Phenylsulfenyl choride: Benzenesulfenyl chloride (8,9); (931-59-9)

Nitromethane: Methane, nitro- (8,9); (75-52-5)

PREPARATION AND USE OF LITHIUM ACETYLIDE:

1-METHYL-2-ETHYNYL-endo-3,3-DIMETHYL-2-NORBORNANOL

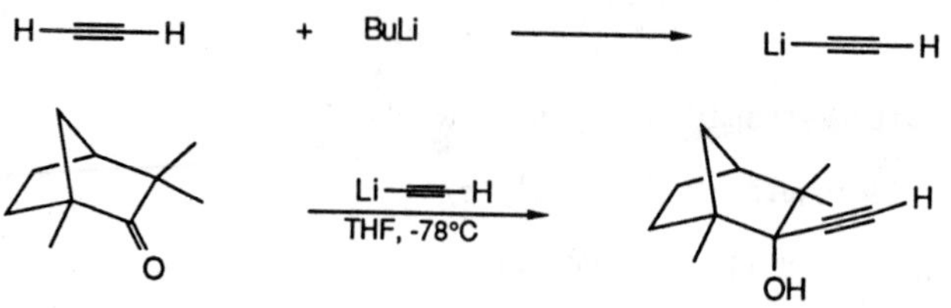

Submitted by M. Mark Midland, Jim I. McLoughlin, and Ralph T. Werley, Jr.[1]
Checked by Matthew J. Sharp and Larry E. Overman.

1. Procedure

An oven-dried (Note 1), 500-mL, septum-capped, round-bottomed flask is flushed with nitrogen, charged with 70 mL of tetrahydrofuran (THF) (Note 2) and cooled to -78°C. After the apparatus has cooled, 157 mL of a 2.1 M solution of butyllithium (0.330 mol) (Note 3) is added using a 50-mL syringe. The contents are mixed by swirling the flask and kept at -78°C until needed. An oven-dried, 2-L, round-bottomed flask, equipped with a magnetic stir bar and capped with a septum, is cooled under a nitrogen purge (Note 4). The flask is charged with 500 mL of THF and cooled to -78°C. A 100-mL graduated cylinder (Note 5) is fitted with a septum in which an 8-mm hole has been bored (Note 6). A 9-mm glass tube which can be connected to either a nitrogen or acetylene line through a three-way valve (Notes 6 and 7) is inserted through the septum of the graduated cylinder. A double-ended needle is used to connect the cylinder to the 2-L reaction flask. After the cylinder

14

and reaction flask have been thoroughly purged with nitrogen, 70 mL of THF is introduced into the cylinder, and the cylinder is cooled to -78°C (Note 8). Acetylene (Notes 7 and 9) is introduced through the 9-mm tube into the bottom of the graduated cylinder at such a rate that 30 mL (Note 10) has been added in 20 min; excess acetylene which does not dissolve in the THF is allowed to flow through the double-ended needle to the 2-L reaction flask. The cold acetylene solution is transferred via the double-ended needle to the 2-L reaction flask. While the solution is cooling in the -78°C bath, nitrogen is blown over the surface to purge completely the system of acetylene (Note 11). The precooled butyllithium/THF solution is then slowly added by a double-ended needle over a period of 1 hr (Note 12). The clear solution of lithium acetylide is stirred for an additional 15 min at -78°C before 48.4 mL of (-)-fenchone (0.300 mol, Note 13) is slowly added via syringe. The solution immediately becomes yellow. After the addition is complete, the cold bath is removed and the mixture is stirred for 3 hr while it warms to room temperature. The flask is opened to the atmosphere and 400 mL of 1.0 M hydrochloric acid is added (Note 14). The quenched reaction is stirred for 20 min and then transferred to a 2-L separatory funnel. The aqueous material is separated and 200 mL of pentane is added. The organic layer is then sequentially washed with 100 mL of 1.0 M hydrochloric acid, 300 mL of water and finally 100 mL of saturated brine. The combined aqueous material is extracted with 200 mL of ether. The organic extracts are combined, dried with magnesium sulfate and filtered. Concentration (rotary evaporation, 40°C under aspirator pressure) provides a thick yellow oil which is distilled to obtain 48.2 g (90%) of the alcohol as a pale yellow oil (bp 51-55°C/0.05 mm) (Note 15).

15

2. Notes

1. All glassware was oven-dried for at least 24 hr at 130°C, assembled hot and cooled under a stream of nitrogen.[2]

2. Tetrahydrofuran, THF, was obtained from Mallinckrodt Inc. and distilled from sodium/benzophenone ketyl immediately prior to use.

3. The checkers used butyllithium purchased from Lithium Alkyls which was standardized by titration with 2,5-dimethoxybenzyl alcohol. The submitters used butyllithium obtained from Aldrich Chemical Company, Inc. as a 1.55 M solution in hexanes which was measured by transfer to a septum-capped 250-mL graduated cylinder with a 15-gauge cannula and nitrogen back pressure.[2a] This solution was then transferred to the flask containing THF in the same manner. The THF must be cooled prior to addition of the butyllithium. Stainless steel double-ended needles of the type used by the submitters are available from Aldrich Chemical Company, Inc. and Ace Glass, Inc.

4. A slight positive pressure of nitrogen was maintained in the reaction vessel throughout the procedure until the reaction was quenched.

5. Graduated cylinders of various sizes with 24/40 standard taper joints are available from Ace Glass, Inc.; appropriate septa were obtained from Aldrich Chemical Company, Inc.

6. Both nitrogen and acetylene are introduced into the graduated cylinder through a 9-mm glass tube approximately 30 cm in length. An 8-mm hole bored through the septum allows for a good seal and for movement of the glass tube in order to adjust the height of the tube from the bottom of the graduated cylinder. A drawing of the apparatus employed is provided in Figure 1.

7.   Acetylene is obtained from Liquid Carbonics and is purified by passage through a -78°C cold trap, a liquid trap (to prevent aspiration of sulfuric acid), a gas washing bottle containing 100 mL of concd sulfuric acid and a calcium chloride drying tube before introduction into the graduated cylinder.[3] A bubbler and a 3-way valve, in that order, are placed before the graduated cylinder. The bubbler with a head of 30 mm of mercury is connected to the system with a T-shaped connecting tube and serves as a relief valve to prevent over pressurization of the acetylene line. The 3-way valve allows nitrogen to be flushed through the graduated cylinder and reaction flask for purging the apparatus and to provide back pressure for the transfer of the acetylene solution. See Figure 1 for a drawing of the reaction apparatus.

Caution: Reactions with acetylene should be carried out in the hood and all lines carrying exhaust from the bubblers must be vented to a fume hood!

8.   The graduated cylinder and its contents are cooled in a wide-mouth Dewar containing a dry ice/acetone slurry.

9. Acetylene is an explosive compound and reacts with metals (e.g., Cu, Ag).[4]

10. The amount of acetylene is approximated by assuming a density of 0.7 g/mL. Approximately 2.5-3.0 equiv of acetylene is used. THF is added to the graduated cylinder so that the final volume of the acetylene solution will be 100 mL. In this case, 30 mL of acetylene is measured to provide approximately 0.85 mol.

11.   Acetylene gas must be removed from above the solution to prevent reaction with the concentrated butyllithium solution entering the flask. Such a reaction in the presence of excess butyllithium will result in the formation of dilithium acetylide on the needle tip. The formation of even small amounts of dilithium acetylide must be avoided; lithium acetylide readily

disproportionates to dilithium acetylide and acetylene upon warming or in the presence of excess butyllithium. The formation of a small amount of dilithium acetylide accelerates the rate of disproportionation (see: Note 12).[5]

12.   Butyllithium must be added slowly to an excess of the efficiently stirred acetylene solution at -78°C. Localized warming of the solution or rapid introduction of butyllithium to produce a local excess of base must be avoided in order to prevent the formation of the unreactive and insoluble dilithium acetylide which will be observed as a cloudy suspension. Reaction of ketones or aldehydes with a cloudy suspension of dilithium acetylide results in substantially lowered yields of the carbinol products.[5]

13.   (-)-Fenchone was obtained from Fluka Chemical Corporation and was used without additional purification.

14.   After the solution is warmed to room temperature, substantial amounts of acetylene and butane may remain in solution. The reaction must be quenched in an efficient fume hood, and the first 10-20 mL of hydrochloric acid solution should be added slowly.

15.   The product displays the following physical and spectral data: $[\alpha]_D^{25}$ +20.4° (CHCl$_3$, c 9); IR (neat film/NaCl plates) cm^{-1}: 3490, 2110, 1460, 1060; ^{1}H NMR (200 MHz, CDCl$_3$) δ: 0.93 (s, 3 H), 1.0-1.15 (m, 2 H), 1.11 (s, 3 H), 1.17 (s, 3 H), 1.20-1.95 (m, 5 H), 1.99 (OH), 2.53 (s, 1 H); ^{13}C NMR (50 MHz, CDCl$_3$) δ: 17.84, 21.54, 25.77, 27.08, 29.81, 40.95, 42.90, 48.40, 53.06, 74.86, 80.45, 85.61; MS (eI, -70 eV): m/z calcd. for C$_{12}$H$_{18}$O: 178.1357, m/z found: 178.1355. The product is a 97:3 mixture of endo:exo addition products based upon capillary GC analysis.   (A Hewlett Packard 5880 Capillary Gas Chromatograph equipped with Supelcowax-10 30-m capillary column available from Supelco, Inc. was used.)

3. Discussion

A general procedure for the preparation and use of monolithium acetylide is described. Monolithium acetylide is a useful reagent for the preparation of a variety of propargyl alcohols and terminal acetylenes.[6] The formation of monolithium acetylide is often complicated by the production of dilithium acetylide.[7] The reagent may be formed in liquid ammonia which serves to stablize the monoanion. Other amines, such as ethylenediamine, may be similarly added as a complexing agent.[8] However, it is often desirable to prepare the more reactive amine-free acetylide species.[5,7b] Dilithium acetylide is an insoluble salt in tetrahydrofuran and is generally unreactive toward ketones, aldehydes, and other electrophiles. The slow addition of a dilute, cooled solution of butyllithium is critical for the reproducible preparation of a clear lithium acetylide solution. Solutions of monolithium acetylide must be kept cold, near -78°C, to prevent disproportionation to dilithium acetylide and acetylene. If the solution is warmed to 0°C, irreversible formation of dilithium acetylide as the white insoluble precipitate occurs.

Lithium acetylide adds in high yield to a variety of ketones and aldehydes (Table I). Typically, 1.1 to 1.2 equiv of lithium acetylide is employed. Sterically-hindered ketones react in higher yield when 2.0 equiv of lithium acetylide is used. Optimum yields are obtained with a concentration of monolithium acetylide of approximately 0.5 M; higher concentrations, approaching 1.0 M, usually result in slightly lowered yields.[5] In all cases the reactions are essentially complete upon warming to room temperature. This method allows for a rapid and convenient preparation of propargyl alcohols. The procedure seems to be generally applicable to a wide variety of ketones and aldehydes.

19

TABLE I

ADDITION OF MONOLITHIUM ACETYLIDE TO ALDEHYDES AND KETONES

Ketone RCOR'		
R	R'	Yield (%)[a]
CH_3	CH_3	94
$CH_3(CH_2)_4$	H	98
CH_3	$(CH_2)_3CH_3$	92
$CH_3CH(CH_3)CH_2$	$CH_3CH(CH_3)CH_2$	75,86[b]
$CH_3CH_2CH(CH_3)$	$CH_3CH_2CH(CH_3)$	89
$(CH_3)_3C$	$(CH_3)_3C$	66,98[b]
$PhCH_2$	CH_3	94
Ph	CH_3	75
Ph	H	93
Ph	Ph	(85)
$PhCH=CH$	H	96
$(CH_3)_2C=CH$	CH_3	86(77)
β-Ionone		93
cyclo-$(CH_2)_4$		94
cyclo-$(CH_2)_5$		95
cyclo-$(CH_2)_6$		90(83)
cyclo-$(CH_2)_7$		86
Norcamphor		97(92)[c]
Cyclohexanecarboxaldehyde		98

[a]By VPC based on RCOR'. Isolated yields are in parentheses. [b]100% Excess
monolithium acetylide was used. [c]The product was >99% 2-ethynyl-endo-2-
norbornanol by VPC and ^{13}C NMR examination.

Figure 1

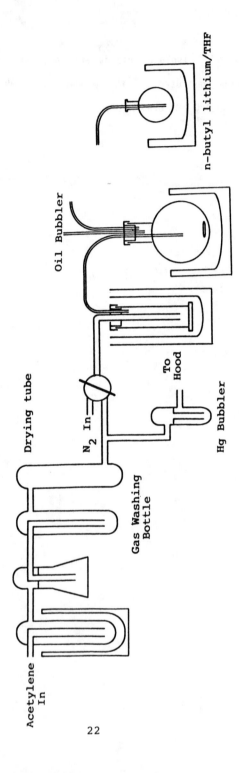

Drying tube

Gas Washing Bottle

N_2 In

Oil Bubbler

To Hood

Hg Bubbler

n-butyl lithium/THF

1. Department of Chemistry, University of California, Riverside, CA 92521.

2. (a) Techniques for handling air sensitive compounds are discussed in: Brown, H. C.; Kramer, G. W.; Levy, A. B.; Midland, M. M. "Organic Synthesis via Boranes"; Wiley: New York, 1975; (b) Technical data for handling air sensitive compounds may also be obtained from Aldrich Chemical Company, Inc.: Lane, C. F.; Kramer, G. W. *Aldrichimica Acta* **1977**, *10*, 11-16.

3. Reichert, J. S.; Nieuwland, J. A. *Org. Synth., Collect. Vol. 1* **1941**, 229.

4. Vogel, A. I.; Furniss, B. J.; Hannaford, A. J.; Rogers, V.; Smith, P. W. G.; Tatchell, A. R. "Textbook of Practical Organic Chemistry", 4th ed.; Longman Inc.: New York, 1978; p. 9.

5. Midland, M. M. *J. Org. Chem.* **1975**, *40*, 2250.

6. Fieser, L. F.; Fieser, M. "Reagents for Organic Synthesis", Wiley: New York, 1967; Vol. 1, pp. 573-574.

7. (a) Corbellini, M.; Turner, L. *Chim. Ind. (Milan)* **1960**, *42*, 251; *Chem. Abstr.* **1960**, *54*, 19250f; (b) Bradsma, L.; Verkruijsse, H. D. "Synthesis of Acetylenes, Allenes and Cumulenes: A Laboratory Manual"; Elsevier: New York, 1981; pp. 9-12.

8. Beumel, O. F., Jr.; Harris, R. J. *J. Org. Chem.* **1963**, *28*, 2775.

Appendix

Chemical Abstracts Nomenclature (Collective Index Number);

(Registry Number)

Monolithium acetylide: Lithium acetylide (8,9); (1111-64-4)

(-)-Fenchone: 2-Norbornanone, 1,3,3-trimethyl- (8); Bicyclo[2.2.1]heptan-2-
one, 1,3,3-trimethyl- (9); (1195-79-5)

ACETYLTRIMETHYLSILANE

(Silane, acetyltrimethyl-)

$$H_2C=CH-OMe \xrightarrow[\text{2. Me}_3\text{SiCl}]{\text{1. LiCMe}_3} \quad H_2C=C(SiMe_3)-OMe \xrightarrow{H_3O^+} \quad CH_3\overset{O}{\underset{\|}{C}}-SiMe_3$$

Submitted by John A. Soderquist.[1]

Checked by Edward D. White III and James D. White.

1. Procedure

Caution! *Persons following these procedures should be thoroughly familiar with the handling of air-sensitive solutions (Note 1).*

A dry (Note 2), 2-L, round-bottomed flask containing a magnetic stirring bar and equipped with a rubber septum inlet is flushed with dry nitrogen and charged with 450 mL of purified tetrahydrofuran (Note 3). After the contents of the flask are cooled using a dry ice/acetone bath, 72 g (1.2 mol) of methyl vinyl ether is distilled into the flask from a commercial cylinder (Note 4). The septum is replaced, under a positive pressure of nitrogen gas, with a 500-mL, pressure-equalizing addition funnel equipped with a rubber septum inlet. The stirred contents of the flask are continuously cooled while 1.0 mol of tert-butyllithium in pentane solution (*Caution!* *Solutions of this reagent are pyrophoric and extreme care should be exercised when carrying out this manipulation!*) (Note 5) is transferred to the addition funnel and subsequently added to the mixture dropwise over ca. 1.5 hr. The resulting yellow slurry is allowed to warm slowly to 0°C over ca. 3 hr (Note 6). With the addition of dry ice to the bath, the near-colorless solution is recooled to ca. -78°C and

25

84.6 g (0.78 mol) of chlorotrimethylsilane (Note 7) is transferred to the addition funnel and subsequently added dropwise to the stirred mixture. The cold bath is removed and, after the mixture has reached room temperature, it is allowed to stir for an additional 1 hr (Note 8). The contents of the flask are carefully poured into a 2-L separatory funnel which contains ca. 400 g of ice and ca. 200 mL of saturated ammonium chloride solution (Note 9). After separation of the aqueous layer, the organic solution is washed with water (12 x 250 mL) (Note 10), dried over anhydrous potassium carbonate, filtered and distilled to give 89-95 g (88-94%) of 1-(methoxyvinyl)trimethylsilane (bp 102-104°C, 760 mm), n_D^{20} 1.4173 (Notes 11 and 12).

A 500-mL, round-bottomed flask is charged with 65 g (0.50 mol) of 1-(methoxyvinyl)trimethylsilane and 300 mL of a 4:1 v/v mixture of acetone and 1.0 M aqueous hydrochloric acid (Note 13). After the pale yellow-green solution is stirred for 1 hr at room temperature, it is transferred to a 1-L separatory funnel and 150 mL each of water and diethyl ether are added. After separation, the aqueous layer is washed with diethyl ether (2 x 50 mL), and these extracts are combined with the organic material. The ethereal solution is washed with water (3 x 300 mL), dried over anhydrous magnesium sulfate, filtered and distilled to give 45-48 g (78-83%) of acetyltrimethylsilane (bp 112°C, 760 mm), n_D^{20} 1.4125 [lit.[2] 1.4113] (Notes 14 and 15).

2. Notes

1. For a detailed description of the general techniques used in the handling of air-sensitive solutions, consult ref. 3. Alternatively, a useful pamphlet describing these techniques is available from Aldrich Chemical Company, Inc., upon request.

26

2. All of the glassware used in the preparation of 1-(methoxyvinyl)-trimethylsilane was dried for at least 4 hr at 110°C, assembled hot and allowed to cool under a nitrogen atmosphere.

3. Reagent-grade tetrahydrofuran (Aldrich Chemical Company, Inc.) was distilled under a nitrogen atmosphere from sodium/benzophenone prior to use.

4. Commercial cylinders of methyl vinyl ether were obtained from Matheson Gas Products, East Rutherford, NJ, and were fitted with a standard needle-valve regulator adapted to connect to a 24 in. 16-gauge syringe needle (Aldrich Chemical Company, Inc.). Unwanted, non-volatile impurities from the cylinder were removed by employing a second flask, which was empty and septum-sealed, between the cylinder and the cold reaction flask. The gaseous reagent is transferred, through the needle, into the empty flask and through a double-ended needle (Aldrich Chemical Company, Inc.) into the reaction flask. The weight of the added methyl vinyl ether was periodically determined from the difference between the initial tare weight of the flask plus the contents and the total weight at each new weighing. However, the amount of this reagent added can be varied by at least +/- 10% without significantly affecting the product yield.

5. This reagent was obtained either from Aldrich Chemical Company, Inc., or Lithium Corporation of America, Bessemer City, NC. A technical data sheet is available from the suppliers. Solutions of ca. 2 M were titrimetrically analyzed for active alkyllithium by the tosylhydrazone method.[4] It is advisable to make certain that the organolithium reagent to be used was prepared in pentane solution. This evaluation can be easily accomplished by the gas chromatographic analysis of the organic layer obtained from the hydrolysis, under a nitrogen atmosphere, of the tert-butyllithium solution to be used. Isobutane and pentane should comprise essentially all of the

volatile material observed. Recently, from the latter supplier, we found that the solvent used was a petroleum distillate which contained some higher boiling components. As a consequence, the distillative isolation of the product was more difficult and the yield was lower (i.e., 80%).

6. The slow warm-up is accomplished conveniently by the removal of the solid dry ice from the cold bath. This process minimizes the formation of acetylenic by-products. The warm-up period can vary with the amount of coolant used. We have consistently obtained good results using sufficient coolant to match the liquid level of the reaction mixture.

7. Chlorotrimethylsilane was purchased from Petrarch, Inc., Levittown, PA and was distilled from calcium hydride prior to use.

8. Lithium chloride precipitates during the warm-up procedure.

9. *CAUTION! The reaction mixture contains low boiling components and care must be exercised to prevent product loss.*

10. This procedure effectively removes the tetrahydrofuran, thus simplifying the distillative isolation of the product. This operation must be done carefully to prevent the loss of product.

11. A Nester-Faust Model NFT-50 annular spinning-band distillation unit was used to obtain the reported product yields in >99% chemical purity by gas chromatographic analysis (Perkin-Elmer Model Sigma 1B Instrument using a 6' x 1/8" 5% SE-30 on silylated Chromosorb W column). Lower product yields (84-86%) of similar chemical purity were obtained using a 200-mm column packed with glass helices.

12. This product gave a satisfactory combustion analysis for $C_6H_{14}OSi$ and exhibited the following spectroscopic data: ^{1}H NMR (CDCl$_3$) δ: 0.09 (s, 9 H), 3.50 (s, 3 H), 4.28 (d, 1 H, J = 2.0), 4.59 (d, 1 H, J = 2.0); ^{13}C NMR (CDCl$_3$) δ: -2.3 (Si-CH$_3$), 54.0 (OCH$_3$), 93.3 (C-2), 170.1 (C-1); IR (film)

cm^{-1}: 1590 (C=C); 1257 (TMS); 1227, 1050 (C=C-OR); MS: m/z 130 (6%); 115 (20%); 89 (47%); 73 (100%); 59 (29%); 44 (11%); 42 (15%).

13. Exposure to direct sunlight was routinely avoided because of the known photochemical reactivity of acylsilanes.[5]

14. Distillation and gas chromatographic analysis of this compound was carried out as described in Note 11.

15. This product gave a satisfactory combustion analysis for $C_5H_{12}OSi$ and exhibited the following spectroscopic data: ^{1}H NMR (CDCl$_3$) δ: 0.09 (s, 9 H), 2.16 (s, 3 H); ^{13}C NMR (CDCl$_3$) δ: -3.5 (Si-CH$_3$), 35.2 (CH$_3$), 246.8 (C=O); IR (film) cm^{-1}: 1645 (C=O); MS: m/z 116 (13%), 101 (11%), 73 (100%), 59 (6%), 44 (34%), 42 (15%). The ultraviolet spectrum of this material in cyclohexane solution exhibits absorbances at 381, 365, 356, and 344 nm with molar extinction coefficients of 91, 123, 97, and 60, respectively. In addition, shorter wave-length shoulders are observed. For a detailed discussion of the spectroscopic properties of acylsilanes, consult ref. 6.

3. Discussion

Acylsilanes have been known since 1957 when Brook described the synthesis of benzoyltriphenylsilane.[7] Their unique reactivity has made them very useful reagents for organic syntheses.[8] The simplest known member of this class of compounds, acetyltrimethylsilane, has been prepared from the oxidation of 1-trimethylsilylethanol,[6] the hydrolysis of 2-methyl-2-trimethylsilyl-1,3-dithiane,[2] the silylation/hydrolysis of N-acetylimidazole,[9] the lithiation/silylation/hydrolysis of ethyl vinyl ether,[10] and the pyrolysis of 2,4,4-trimethyl-2-trimethylsilyl-1,3-oxathiolane 3,3-dioxide,[11] and the silylation of acid chlorides.[12]

The present preparation[13] utilizes the simple deprotonation of methyl vinyl ether first reported by Baldwin and co-workers to obtain 1-(methoxyvinyl)lithium,[14] which functions as a very useful reagent for nucleophilic acylation.[15] After detailed studies of the processes involved, this approach has been applied to the syntheses of a number of acyl derivatives of silicon, germanium, and tin.[16] This procedure also overcomes the hydrolysis problems encountered in a previous study.[10] The vinyl ether approach to such acylmetalloids has been demonstrated to provide access to systems which cannot be prepared using other acyl anion equivalents.[17] In the present case, the simple two-step process from commercially-available reagents gives the highest reported overall yield of pure acetyltrimethylsilane from chlorotrimethylsilane (69-78%). Moreover, the reaction sequence can be scaled up or down without encountering difficulties.

1. Chemistry Department, University of Puerto Rico, Rio Piedras, PR 00931.

2. Brook, A. G.; Duff, J. M.; Jones, P. F.; Davis, N. R. *J. Am. Chem. Soc.* **1967**, *89*, 431.

3. Brown, H. C.; Kramer, G. W.; Levy, A. B.; Midland, M. M. "Organic Syntheses via Boranes;" Wiley-Interscience: New York, 1975; Chapter 9.

4. Lipton, M. F.; Sorenson, C. M.; Sadler, A. C.; Shapiro, R. H. *J. Organomet. Chem.* **1980**, *186*, 155.

5. Brook, A. G. *Acc. Chem. Res.* **1974**, *7*, 77; Bourque, R. A.; Davis, P. D.; Dalton, J. C. *J. Am. Chem. Soc.* **1981**, *103*, 697.

6. Brook, A. G. *Adv. Organomet. Chem.* **1968**, *7*, 95; Dexheimer, E. M.; Buell, G. R.; LeCroix, C. *Spectrosc. Lett.* **1978**, *11*, 751; Bernardi, F.; Lunazzi, L.; Ricci, A.; Seconi, G.; Tonachini, G. *Tetrahedron* **1986**, *42*, 3607.

7. Brook, A. G. *J. Am. Chem. Soc.* **1957**, *79*, 4373.

8. Colvin, E. W. "Silicon in Organic Synthesis;" Butterworths: London, 1981; Weber, W. P. "Silicon Reagents for Organic Synthesis;" Springer-Verlag: Berlin, 1983.

9. Bourgeois, P.; Dunogues, J.; Duffaut, N.; Lapouyade, P. *J. Organomet. Chem.* **1974**, *80*, C25.

10. Dexheimer, E. M.; Spialter, L. *J. Organomet. Chem.* **1976**, *107*, 229.

11. Gokel, G. W.; Gerdes, H. M.; Miles, D. E.; Hufnal, J. M.; Zerby, G. A. *Tetrahedron Lett.* **1979**, 3375.

12. Capperucci, A.; Degl'Innocenti, A.; Faggi, C.; Ricci, A.; Dembech, P.; Seconi, G. *J. Org. Chem.* **1988**, *53*, 3612.

13. Soderquist, J. A.; Hsu, G. J.-H. *Organometallics* **1982**, *1*, 830; Soderquist, J. A.; Hassner, A. *J. Organomet Chem.* **1978**, *156*, C12.

14. Baldwin, J. E.; Höfle, G. A.; Lever, Jr., O. W. *J. Am. Chem. Soc.* **1974**, *96*, 7125.

15. Lever, Jr., O. W. *Tetrahedron* **1976**, *32*, 1943.

16. Soderquist, J. A.; Hassner, A. *J. Am. Chem. Soc.* **1980**, *102*, 1577.

17. Soderquist, J. A.; Hassner, A. *J. Org. Chem.* **1980**, *45*, 541. See also ref. 16.

Appendix
Chemical Abstracts Nomenclature (Collective Index Number); (Registry Number)

Acetyltrimethylsilane: Silane, acetyltrimethyl- (8,9); (13411-48-8)

Methyl vinyl ether: Ether, methyl vinyl (8); Ethene, methoxy- (9); (107-25-5)

Chlorotrimethylsilane: Silane, chlorotrimethyl- (8,9); (75-77-4)

1-(Methoxyvinyl)trimethylsilane: Silane, (1-methoxyethenyl)trimethyl- (10); (79678-01-6)

3-BUTYLCYCLOBUTENONE

(2-Cyclobuten-1-one, 3-butyl-)

$$Bu-C{\equiv}CH \quad \xrightarrow[\text{Et}_2\text{O}]{\substack{\text{CCl}_3\text{COCl} \\ \text{Zn(Cu), DME}}} \quad \text{[structure]} \quad \xrightarrow[\text{EtOH}]{\substack{\text{Zn} \\ \text{TMEDA-AcOH}}} \quad \text{[structure]}$$

Submitted by Rick L. Danheiser,[1a] Selvaraj Savariar, and Don D. Cha.[1b]
Checked by Reinhard Kratzberg and Ekkehard Winterfeldt.

1. Procedure

3-Butyl-4,4-dichlorocyclobutenone. A 1-L, three-necked, round-bottomed flask is equipped with a magnetic stirring bar, two glass stoppers, and a 250-mL pressure-equalizing addition funnel fitted with a nitrogen inlet adapter (Note 1). The flask is charged with 39.23 g (0.60 mol) of zinc-copper couple (Note 2), 400 mL of diethyl ether (Note 3), and 23.0 mL (0.20 mol) of 1-hexyne (Note 4). The dropping funnel is charged with a solution of 44.6 mL (0.40 mol) of trichloroacetyl chloride (Note 5) in 125 mL of dimethoxyethane (Note 6), and this solution is then added dropwise to the reaction mixture over 1 hr. After 18 hr, the resulting brown mixture is filtered through a sintered-glass Büchner funnel, and the black solid which is separated is thoroughly washed with 200 mL of hexane. The filtrate is washed successively with 200 mL each of ice-cold 0.5 N hydrochloric acid, ice-cold 5% sodium hydroxide solution, and saturated sodium chloride solution, dried over anhydrous magnesium sulfate, and then concentrated at reduced pressure using a rotary evaporator. The residual brown oil is transferred to a 100-mL, round-bottomed

32

flask and distilled through a 10-cm Vigreux column to afford 29.5-30.0 g (76-78%) of 3-butyl-4,4-dichlorocyclobutenone as a colorless liquid, bp 68.5-70°C (0.3 mm) (Note 7).

3-Butylcyclobutenone. A 1-L, three-necked, round-bottomed flask is equipped with a magnetic stirring bar, two glass stoppers, and a 125-mL pressure-equalizing dropping funnel fitted with a nitrogen inlet adapter (Note 1). The flask is charged with 52.6 g (0.805 mol) of zinc dust (Note 8), 121 mL of tetramethylethylenediamine (Note 9), and 270 mL of absolute ethanol, and cooled with an ice bath while 46 mL of glacial acetic acid is added dropwise over 5 min. The reaction mixture is maintained at 0°C while a solution of 26.59 g (0.138 mol) of 3-butyl-4,4-dichlorocyclobutenone in 27 mL of absolute ethanol is added over 10 min via the dropping funnel. After 15 min the ice bath is removed, and the reaction mixture is stirred for 2.5 hr and then filtered through a sintered glass Büchner funnel with the aid of 1.5 L of a 1:1 mixture of diethyl ether and pentane. The filtrate is washed successively with 500 mL of 1 N hydrochloric acid, 500 mL of water, 800 mL of saturated sodium bicarbonate solution, and 500 mL of saturated sodium chloride solution, dried over anhydrous magnesium sulfate, filtered, and concentrated at reduced pressure using a rotary evaporator. The residual yellow oil is transferred to a 100-mL, round-bottomed flask fitted with a short-path distillation head (Note 10) and a pear-shaped receiver flask which is cooled below -75°C with a dry ice-acetone bath. Distillation at 0.001 mm (bath temperature 50-70°C) (Note 11) provides 3-butylcyclobutenone (13.4-16.2 g, 78-86% yield) as a clear, colorless liquid, bp 33°C (0.001 mm) (Note 12).

2. Notes

1. The glass components of the apparatus are dried overnight in a 120°C oven and then assembled and maintained under an atmosphere of nitrogen during the course of the reaction.

2. To a stirred mixture of 65.38 g (1.0 mol) of zinc dust (purchased from Mallinckrodt Inc.) and 100 mL of water in a 1-L Erlenmeyer flask is added at 30-sec intervals two solutions of 7.6 g of copper sulfate, ($CuSO_4 \cdot 5H_2O$) in 50 mL of water. After 1 min the mixture is filtered through a sintered-glass Büchner funnel and the zinc-copper couple is washed with two 50-mL portions of water, two 50-mL portions of acetone, and 50 mL of diethyl ether. The resulting dark gray powder is dried at 100°C at 1 mm for 4 hr and then stored under nitrogen.

3. Diethyl ether was distilled from sodium benzophenone ketyl immediately before use.

4. 1-Hexyne was obtained from Aldrich Chemical Company, Inc., and distilled from calcium hydride before use.

5. Trichloroacetyl chloride was purchased from Fluka Chemical Corporation and distilled before use.

6. Dimethoxyethane was obtained from Aldrich Chemical Company, Inc., and distilled from sodium benzophenone ketyl immediately before use.

7. The product exhibits the following spectral properties: IR (film) cm^{-1}: 2970, 2940, 2880, 1800, 1585, 1475, 1415, 1390, 1260, 1210, 1045, 850, and 630; UV (isooctane) nm max (ε): 306 (33) and 215 (9530); ^{1}H NMR (300 MHz, CDCl$_3$) δ: 0.96 (t, 3 H, J = 7), 1.45 (apparent sextet, 2 H, J = 7), 1.72 (apparent quintet, 2 H, J = 7), 2.69 (dt, 2 H, J = 2, 7), and 6.20 (t, 1 H, J = 2); ^{13}C NMR (75.4 MHz, CDCl$_3$) δ: 13.5, 22.3, 25.9, 27.6, 91.9, 135.6, 179.2, and 186.1.

34

8. Zinc dust was purchased from Mallinckrodt Inc. and used without further purification.

9. Tetramethylethylenediamine (TMEDA) was obtained from Aldrich Chemical Company, Inc., and distilled from calcium hydride before use.

10. The delivery tube of the short-path distillation head is fitted with a 5-cm length of Teflon tubing which extends nearly to the bottom of the receiving flask.

11. Substantial decomposition of the product was observed if the bath temperature was allowed to exceed 80°C or if distillation was attempted using a short Vigreux column.

12. 3-Butylcyclobutenone displays the following spectral properties: IR (film) cm^{-1}: 2970, 2940, 2860, 1770, 1590, 1470, 1420, 1385, 1310, 1280, 1220, 1180, 1100, 1030, 990, 925, and 850; UV (isooctane) nm max (ε): 306 (36) and 220 (6860); ^{1}H NMR (300 MHz, CDCl$_3$) δ: 0.93 (t, 3 H, J = 7), 1.38 (apparent sextet, 2 H, J = 7), 1.58 (apparent quintet, 2 H, J = 7), 2.54 (t, 2 H, J = 7), 3.12 (s, 2 H), and 5.86 (s, 1 H); ^{13}C NMR (75.4 MHz, CDCl$_3$) δ: 13.5, 22.2, 27.9, 31.6, 50.5, 133.8, 181.2, and 187.6.

3. Discussion

The procedure described above illustrates a general two-step method for the preparation of 3-substituted and 2,3-disubstituted cyclobutenones.[2,3] The first step in the procedure involves the [2+2] cycloaddition of an acetylene with dichloroketene, which is best carried out using a modification of the method originally reported by Hassner and Dillon.[4] Under these conditions dichloroketene combines smoothly with a variety of alkyne derivatives in contrast to ketene itself which fails to add to unactivated acetylenes. The

35

4,4-dichlorocyclobutenone cycloadducts produced in the first step are then subjected to reductive dechlorination with zinc dust to afford the desired cyclobutenones. Although it has previously been reported that this reductive dechlorination reaction cannot be accomplished reliably,[4] recent studies[2,3] have shown that under carefully controlled conditions the desired transformation is in fact a feasible and efficient process. As summarized in Table I, this strategy is applicable to the synthesis of a variety of substituted cyclobutenones.

The preparation of 3-butyl-4,4-dichlorocyclobutenone described here represents an optimized procedure for the [2+2] cycloaddition of dichloroketene with acetylenes. Previously there has been some disagreement over the efficacy of phosphorus oxychloride ($POCl_3$) as a sequestering agent for the zinc chloride generated in the course of the reaction. Hassner and Dillon[4] have reported that the inclusion of phosphorus oxychloride is crucial to the success of these reactions, and that in its absence tarry products are produced which are difficult to purify. Dreiding and co-workers, however, report that these conditions lead to the production of the desired cycloadducts contaminated with significant amounts of the corresponding 2,4-dichloro isomers. Dreiding consequently suggests that these cycloadditions are best carried out in the absence of phosphorus oxychloride employing short reaction times of less than 15 min.

We have found that the method used to prepare the zinc-copper couple is an important variable in determining the efficiency and rate of these reactions. Optimal results are achieved using a couple prepared by brief (2 min) exposure of commercial zinc dust to twice the amount (0.06 equiv) of copper sulfate employed in the previous studies.[3,4] We have also found that the cycloadditions proceed with equal efficiency and more conveniently by employing dimethoxyethane[5] in place of phosphorus oxychloride as a zinc chloride sequestering agent. Under these optimized conditions a variety of terminal acetylenes are smoothly converted to 4,4-dichlorocyclobutenones in high yield. Although cycloadditions involving disubstituted acetylenes lead to the desired 4,4-dichlorocyclobutenones contaminated with varying amounts of isomeric 2,4-dichloro isomers, the formation of these by-products can be suppressed (to less than 10%) simply by carrying out the cycloaddition at temperatures between 10°C and 15°C.

Under conventional dechlorination conditions (20 equiv of zinc dust, acetic acid, 25°C or 50°C) the reduction of 4,4-dichlorocyclobutenones affords complex mixtures of products which include the desired cyclobutenones as well as significant amounts of partially reduced byproducts. We have found that the desired transformation can be accomplished cleanly provided that the reduction is carried out at room temperature in alcoholic solvents (preferably ethanol) in the presence of 5 equiv each of acetic acid and a tertiary amine (preferably tetramethylethylenediamine). Zinc dust has proven to be somewhat superior to zinc-copper couple for this reduction. The desired cyclobutenones are obtained in somewhat higher yield using this procedure as compared to the related conditions reported by Dreiding [Zn(Cu), 4:1 AcOH-pyridine, 50-60°C] for the same transformation.[3]

The most efficient synthetic route to 3-butylcyclobutenone reported previously[6] involves the addition of butylmagnesium bromide[7] or butyllithium[8] to 3-ethoxycyclobutenone followed by acid hydrolysis. 3-Ethoxycyclobutenone is itself available in modest yield via the addition of ketene to ethoxyacetylene. This procedure provides 3-butylcyclobutenone in only 20% overall yield and requires the use of a ketene generator and the rather unstable ethoxyacetylene as starting material.

37

Cyclobutenones serve as versatile intermediates for the preparation of α,β-butenolides,[9] cyclopentenones,[9] and a variety of substituted cyclobutane derivatives. We have shown that cyclobutenones also function as four-carbon annulation components in routes to eight-membered carbocycles[8] and highly substituted aromatic compounds.[10]

1. (a) Department of Chemistry, Massachusetts Institute of Technology, Cambridge, MA 02139; (b) Berlex Laboratories Predoctoral Fellow, 1986-1987.

2. Danheiser, R. L.; Savariar, S. *Tetrahedron Lett.* **1987**, *28*, 3299.

3. For a closely related method, see Ammann, A. A.; Rey M.; Dreiding, A. S *Helv. Chim. Acta* **1987**, *70*, 321.

4. Hassner, A.; Dillon, J. L., Jr. *J. Org. Chem.* **1983**, *48*, 3382.

5. Johnston, B. D.; Czyzewska, E.; Oehlschlager, A. C. *J. Org. Chem.* **1987**, *52*, 3693.

6. Several multistep routes to 3-butylcyclobutenone have also been reported: (a) Corbel, B.; Decesare, J. M.; Durst, T. *Can. J. Chem.* **1978**, *56*, 505; (b) Takeda, T.; Tsuchida, T.; Ando, K.; Fujiwara, T. *Chem. Lett.* **1983**, 549.

7. Wasserman, H. H.; Piper, J. U.; Dehmlow, E. V. *J. Org. Chem.* **1973**, *38*, 1451.

8. Danheiser, R. L.; Gee, S. K.; Sard, H. *J. Am. Chem. Soc.* **1982**, *104*, 7670.

9. Schmit, C.; Sahraoui-Taleb, S.; Differding, E.; Dehasse-De Lombaert, C. G.; Ghosez, L. *Tetrahedron Lett.* **1984**, *25*, 5043.

10. (a) Danheiser, R. L.; Gee, S. K. *J. Org. Chem.* **1984**, *49*, 1672; (b) Danheiser, R. L.; Gee, S. K.; Perez, J. J. *J. Am. Chem. Soc.* **1986**, *108*, 806.

Table. Synthesis of Cyclobutenones from Alkynes

Case	Alkyne	Cyclobutenone	Isolated Yield (%)
1	Bu-C≡CH		70
2	tert-Bu-C≡CH		54
3	AcO(CH$_2$)$_3$-C≡CH		56
4	Ph-C≡CH		66-77
5	Cyclododecyne		75
6	Pr-C≡C-Pr		73

Appendix

Chemical Abstracts Nomenclature (Collective Index Number)

(Registry Number)

3-Butylcyclobutenone: 2-Cyclobuten-1-one, 3-butyl- (9); (38425-48-8)

3-Butyl-4,4-dichlorocyclobutenone: 2-Cyclobuten-1-one, 3-butyl-4,4-dichloro- (10); (72284-70-9)

1-Hexyne (8,9); (693-02-7)

Trichloroacetyl chloride: Acetyl chloride, trichloro- (8,9); (76-02-8)

Tetramethylethylenediamine: Ethylenediamine, N,N,N',N'-tetramethyl- (8); 1,2-Ethanediamine, N,N,N',N'-tetramethyl- (9); (110-18-9)

VICINAL DICARBOXYLATION OF AN ALKENE:

cis-1-METHYLCYCLOPENTANE-1,2-DICARBOXYLIC ACID

(1,2-Cyclopentanedicarboxylic acid, 1-methyl-, cis-(±)-)

A.

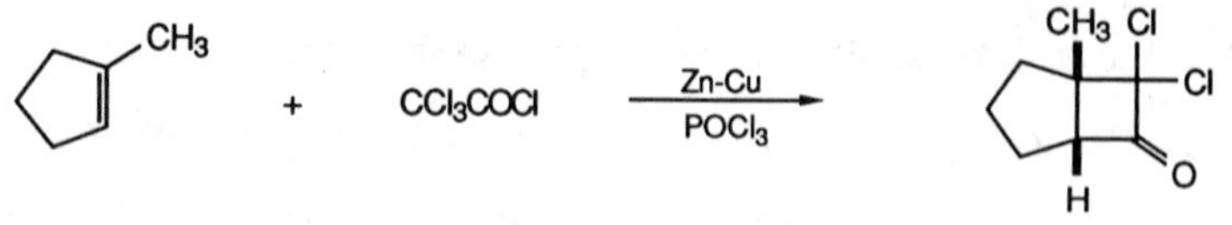

B.

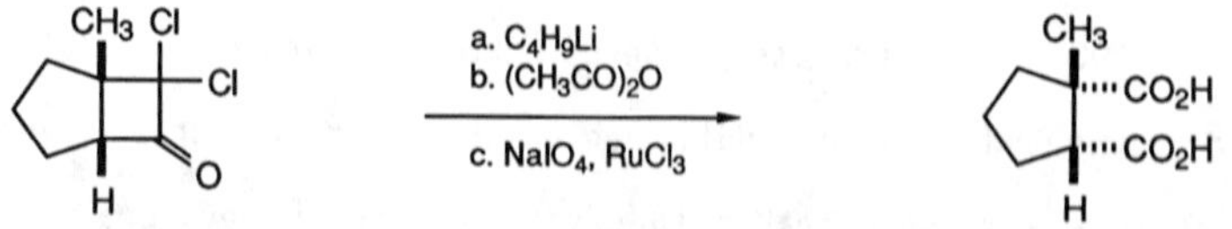

Submitted by Jean-Pierre Deprés and Andrew E. Greene.[1]

Checked by Scott K. Thompson, Gregory A. Slough, and Clayton H. Heathcock.

1. Procedure

A. 7,7-Dichloro-1-methylbicyclo[3.2.0]heptan-6-one. A 500-mL, two-necked, round-bottomed flask is equipped with a Teflon-covered magnetic stirring bar, a 250-mL pressure-equalizing addition funnel topped with a gas inlet, and a condenser connected to a Nujol-filled bubbler (Note 1). The system is flushed with nitrogen (Note 2). The flask is then charged with 10.0 g (ca. 150 mmol) of zinc-copper couple (Note 3), 200 mL of anhydrous ether (Note 4), and 10.5 mL (8.2 g, 100 mmol) of 1-methyl-1-cyclopentene (Note 5) and the addition funnel is filled with a solution of 13.4 mL (21.8 g, 120 mmol) of trichloroacetyl chloride (Note 5) and 11.2 mL (18.4 g, 120 mmol) of phosphorus oxychloride (Note 6) in 100 mL of anhydrous ether. Magnetic

41

stirring is begun and the solution is added dropwise over 1 hr to the reaction flask at room temperature. After being stirred for an additional 14 hr, the reaction mixture is filtered under water pump pressure through 30 g of filter aid, which is then washed with 120 mL of ether. The filtrate is concentrated to ca. 100-120 mL, treated with 400 mL of hexane, and then briefly stirred to precipitate the zinc chloride. The supernatant solution is transferred to a separatory funnel and the viscous residue is washed with two 75-mL portions of 3:1 hexane-ether. The combined solution is washed successively with 200 mL of cold water, 200 mL of saturated aqueous sodium bicarbonate solution, and 2 x 50 mL of saturated aqueous sodium chloride solution, dried over anhydrous sodium sulfate, and concentrated to dryness by rotary evaporation at 25°C to give 17.0-17.8 g of a brown oil. Vacuum distillation of this material without fractionation provides 14.9-16.0 g (77-83%) of 7,7-dichloro-1-methylbicyclo-[3.2.0]heptan-6-one as a clear, light yellow oil, bp 38°C (0.2 mm), n_D^{20} 1.4970 (Note 7).

B.   *cis-1-Methylcyclopentane-1,2-dicarboxylic acid.*   A 1-L, one-necked, round-bottomed flask (Note 1) equipped with a Teflon-covered magnetic stirring bar is flushed with nitrogen and then charged with 300 mL of dry tetrahydrofuran (Note 4) and 14.5 g (75 mmol) of 7,7-dichloro-1-methylbicyclo-[3.2.0]heptan-6-one.   The flask is capped with a septum and connected to a Nujol bubbler and to a nitrogen line by means of syringe needles (Note 2).   To the stirred solution cooled in a dry ice-acetone bath is added by syringe over 5 min 33.2 mL (83 mmol) of a 2.50 M solution of butyllithium in hexane (Note 8).   After being stirred for 15 min with continued cooling, the reaction mixture is treated with 14.2 mL (150 mmol) of acetic anhydride all at once (Note 9).   The cooling bath is removed and the reaction mixture is allowed to warm to room temperature and then stirred for an additional 1 hr.   Most of the

42

solvent and excess acetic anhydride are directly removed with a rotary evaporator at 25°C under water pump pressure (Note 10). The resulting solid residue is further dried for 15 to 30 min at 4 mm and then dissolved in a mixture of 100 mL of acetonitrile, 100 mL of carbon tetrachloride, and 150 mL of distilled water. The mixture is cooled in an ice bath and, with efficient stirring, treated with 40.1 g (187 mmol) of sodium periodate and 346 mg (1.5 mmol) of ruthenium(III) chloride hydrate (Note 11). After 15 min, the cooling bath is removed (Note 12) and stirring is continued for 5 hr, whereupon the thick mixture is treated with 200 mL of 10% aqueous sodium hydroxide solution and then extracted in a separatory funnel with 500 mL of 1:1 ether-hexane (Note 13). The phases are separated and to the aqueous phase is added 900 mL of 2:1 ether-ethyl acetate followed by a 2 N aqueous hydrochloric acid solution until a pH of 2 to 3 is obtained (Note 14). After being vigorously agitated, the phases are separated and the organic phase is washed successively with solutions of 3% aqueous sodium thiosulfate (Note 15) and saturated aqueous sodium chloride. All aqueous phases are mixed and, at pH 2 to 3, extracted with 1 L of 3:2 ether-ethyl acetate, which is then washed as before. The ether-ethyl acetate solutions are combined and dried over anhydrous sodium sulfate and the solvents are removed by rotary evaporation to leave a light yellow solid, mp 123-126°C. Trituration of this material with 1:1 ethyl acetate-petroleum ether (Note 16) gives 7.9-8.0 g (61-62%) of cis-1-methylcyclopentane-1,2-dicarboxylic acid as a white solid, mp 123-124.5°C (Notes 16, 17, 18, and 19).

2. Notes

1. All glassware was dried overnight in an oven at 115°C and allowed to cool in a desiccator.

2. A slight positive pressure of nitrogen is maintained throughout the reaction.

3. A literature procedure[2] for the preparation of the zinc-copper couple was followed except for the use of slightly more (28%) than the indicated amount of copper sulfate. The checkers found that the kind of zinc used is critical. Zinc dust, 325 mesh, from Aldrich Chemical Company, Inc. [catalog number 20,998-8] gave 7,7-dichloro-1-methylbicyclo[3.2.0]heptan-6-one in 80-89% yield. Zinc metal (dust) from Fisher Scientific Company (Lot 874394) gave the dichloro ketone in yields of 37-61% (five trials). The Fisher zinc was of unknown mesh, but was much more finely-divided than the Aldrich Chemical Company, Inc. zinc.

4. Ether and tetrahydrofuran were distilled from the sodium ketyl of benzophenone.

5. 1-Methyl-1-cyclopentene (96% pure) and trichloroacetyl chloride (99% pure) were purchased from the Aldrich Chemical Company, Inc. The trichloroacetyl chloride was distilled prior to use.

6. Phosphorus oxychloride (99% pure) was obtained from the Aldrich Chemical Company, Inc. and distilled from potassium carbonate prior to use.

7. This material was found to darken with time. Its spectral properties are the following: IR (film) cm^{-1}: 1805; ^1H NMR (CDCl$_3$, 80 MHz) δ: 1.57 (s, 3 H), 1.5-2.5 (m, 6 H), 3.50 (m, 1 H). These values are in accord with those in the literature.[3]

8. The solution of butyllithium in hexane was purchased from the Aldrich Chemical Company, Inc. and standardized with menthol and phenanthroline[4] before use.

9. Acetic anhydride was distilled prior to use.

10. A trap is used between the flask and the rotary evaporator as a precaution against possible bumping during the evaporation.

11. Sodium periodate was obtained from the Aldrich Chemical Company, Inc. and ruthenium(III) chloride hydrate (5-10% water) was purchased from Fluka. The more expensive periodic acid can replace sodium periodate; however, ruthenium(III) chloride appears to be somewhat more efficient than ruthenium(IV) oxide.

12. Should a noticeably exothermic reaction ensue, the cooling bath is replaced for a few minutes.

13. At this point the checkers filtered the mixture through a pad of 30 g of Celite to remove the green precipitate. This filtration reduces the problem of emulsions and clogging of the separatory funnel during subsequent extractions.

14. Iodine formation becomes substantial at lower pH.

15. Normally 50-100 mL of this solution is required.

16. This is done first with 20 mL and then with 8 mL. Product loss is minimized by storing the trituration flask overnight at -25°C prior to removal of the supernatant solution. On evaporation, the supernatant solution affords an oil containing the diacid and a small amount of the corresponding anhydride. Treatment of this oil with 10% aqueous NaOH at room temperature overnight and then processing the solution as before yields an additional 0.4 g (3%) of the diacid, mp 126-127°C.

17.   The submitters report a crude yield of 11.5 g (89%) and a recrystallized yield of 9.0 g (70%), mp 128-129°C.

18.   Melting points from 123 to 129°C have been reported for this compound.[5]   Its spectral properties are the following: IR (Nujol) cm^{-1}: 2720, 2630, 1690; ^{1}H NMR ($CDCl_3$, 80 MHz) δ: 1.44 (s, 3 H), 1.5-2.5 (m, 6 H), 2.72 (pseudo t, 1 H, J = 8), 10.8 (br s, 2 H).

19.   Gas chromatographic analysis (10% Carbowax 20 M on 80-100 mesh Chromosorb W, 2.5 m x 2 mm, column temperature 180°C, injection temperature 230°C, flow rate 10 mL/min, retention time 10 min) of the corresponding dimethyl ester, formed with ethereal diazomethane, indicated a purity of greater than 99%.

3. Discussion

This procedure serves to illustrate a relatively inexpensive, two-pot, stereoselective method for effecting vicinal dicarboxylation of alkenes that is more generally applicable and higher-yielding than the palladium-catalyzed carbonylation reaction and other more circuitous procedures.[6] Part A of this procedure is a slight modification of the dichloroketene-olefin cycloaddition method previously described by Krepski and Hassner.[2]   Part B makes use of Sharpless and co-workers' ruthenium(III) chloride-catalyzed oxidation process[7] for the cleavage of the β-chloro enol acetate, which is formed on trapping the β-chloro enolate intermediate with acetic anhydride.   A slightly different, non-optimized version of this procedure has been used[6,8] for the vicinal dicarboxylation of 1-decene, cis- and trans-2-butene, 2-methyl-2-butene, 2,3-dimethyl-2-butene, 1-methyl-1-cyclohexene, 1,6-dimethyl-1-cyclohexene, and 5α-cholest-2-ene with overall yields of 52-83%.

cis-1-Methylcyclopentane-1,2-dicarboxylic acid has been previously prepared by a variety of methods: by oxidative cleavage (HNO_3) of cyclobutanone[5a] and cyclopentanone[5b] precursors, through saponification and oxidation ($KMnO_4$) of a γ-butyrolactone intermediate,[5c] and by anhydride formation and then hydrolysis starting from mixtures of the cis and trans diacids (obtained in ca. 5 steps).[5d,e] Compared with these methods, the cycloaddition-cleavage procedure is much more efficient and practical. It requires only readily available reagents and easily affords, without any chromatographic separations, a product of high purity.

1. Laboratoire d'Etudes Dynamiques et Structurales de la Sélectivité (LEDSS), Université Joseph Fourier de Grenoble, Bât. 52 Chimie Recherche BP. 68, 38402 Saint Martin d'Hères Cedex, France. This work was supported by CNRS (UA 332).

2. Krepski, L. R.; Hassner, A. *J. Org. Chem.* **1978**, *43*, 2879-2882.

3. Jeffs, P. W.; Molina, G.; Cass, M. W.; Cortese, N. A. *J. Org. Chem.* **1982**, *47*, 3871-3875.

4. Watson, S. C.; Eastham, J. F. *J. Organomet. Chem.* **1967**, *9*, 165-168.

5. (a) Wiberg, K. B.; Klein, G. W. *Tetrahedron Lett.* **1963**, 1043-1045; (b) Bagchi, P.; Banerjee, D. K. *J. Indian. Chem. Soc.* **1947**, *24*, 12-14; (c) Julia, M.; Salard, J. M.; Chottard, J. C. *Bull. Soc. Chim. Fr.* **1973**, 2478-2482; (d) Dutta, P. C. *J. Indian. Chem. Soc.* **1940**, *17*, 611-618; (e) Bachmann, W. E.; Struve, W. S. *J. Am. Chem. Soc.* **1941**, *63*, 1262-1265.

6. See: Deprés, J.-P.; Coelho, F.; Greene, A. E. *J. Org. Chem.* **1985**, *50*, 1972-1973, and references cited therein.

7. Carlsen, P. H. J.; Katsuki, T.; Martin, V. S.; Sharpless, K. B. *J. Org. Chem.* **1981**, *46*, 3936-3938.

8. Greene, A. E.; Deprés, J.-P.; Coelho, F.; Brocksom, T. J. *J. Org. Chem.* **1985**, *50*, 3943-3945.

Appendix

Chemical Abstracts Nomenclature (Collective Index Number);

(Registry Number)

cis-1-Methylcyclopentane-1,2-dicarboxylic acid: 1,2-Cyclopentanedicarboxylic acid, 1-methyl-, cis-(±)- (10); (70433-31-7)

7,7-Dichloro-1-methylbicyclo[3.2.0]heptan-6-one: Bicyclo[3.2.0]heptan-6-one, 7,7-dichloro-1-methyl- (9); (51284-43-6)

1-Methyl-1-cyclopentene: Cyclopentene, 1-methyl- (8,9); (693-89-0)

Trichloroacetyl chloride: Acetyl chloride, trichloro- (8,9); (76-02-8)

Sodium periodate: Periodic acid, sodium salt (8,9); (7790-28-5)

Ruthenium(III) chloride hydrate: Ruthenium chloride, hydrate (8,9); (14898-67-0)

ENANTIOSELECTIVE OXIDATION OF A SULFIDE:

(S)-(-)-METHYL p-TOLYL SULFOXIDE

(Benzene, 1-methyl-4-(methylsulfinyl)-, (S)-)

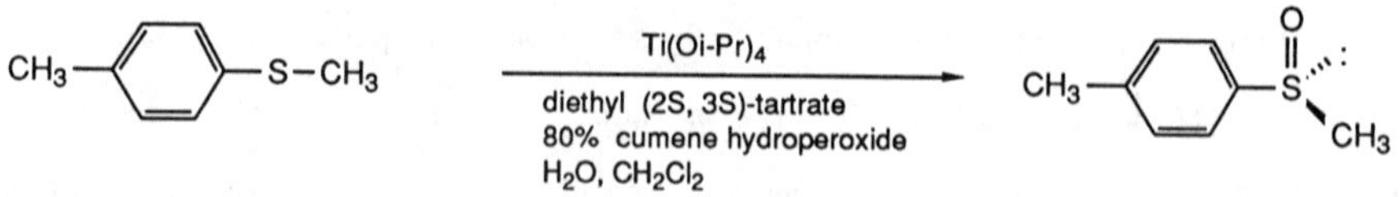

CH_3—⟨benzene ring⟩—S—CH_3 →
[reagents over arrow:]
Ti(Oi-Pr)$_4$
diethyl (2S, 3S)-tartrate
80% cumene hydroperoxide
H_2O, CH_2Cl_2
→ CH_3—⟨benzene ring⟩—S(=O)—CH_3

Submitted by S. H. Zhao, O. Samuel, and H. B. Kagan.[1]

Checked by Carl A. Busacca and Albert I. Meyers.

1. Procedure

Into a 1-L flask containing 125 mL of methylene chloride (Note 1) and a magnetic stirring bar (4-cm length) is added at room temperature (20°C) 5.35 mL, 6.19 g (0.030 mol) of (S,S)-(-)-diethyl tartrate (DET) (Note 2) by means of a 10-mL syringe. The flask is stoppered with a septum cap and purged with argon. Titanium(IV) isopropoxide, Ti(OiPr)$_4$, (4.48 mL, 0.015 mol) (Note 3) is introduced through the septum via a 10-mL syringe. The stirred limpid solution immediately turns yellow. After a few minutes distilled water (0.27 mL, 0.015 mol) is added dropwise using a calibrated syringe. Strong stirring is maintained until there is total dissolution of water and formation of a pale yellow solution (after 25 min). Methyl p-tolyl sulfide (4.09 mL, 4.20 g, 0.030 mol) (Note 4) dissolved in 5 mL of methylene chloride is introduced with a 10-mL syringe which is then rinsed with 5 mL of methylene chloride. The flask is cooled (-30°C) with an acetone-dry ice bath while it is stirred for

49

40 min. At this point 5.54 mL of 80% cumene hydroperoxide (Note 5) (0.030 mol, 5.70 g) is added dropwise from a 10-mL syringe, with stirring, during 5 min. The reaction flask is kept in a freezer (-23°C) overnight (15 hr) (Note 6). Hydrolysis is then effected by adding 5.05 mL of water followed by vigorous stirring for 90 min at room temperature (20°C).

A large sintered-glass funnel (9-cm diameter, porosity grade 2) is partially filled with Celite (Celite height: 2.5 cm) and then impregnated with methylene chloride. The suspension resulting from hydrolysis is poured in portions onto the Celite under suction by a water pump. The Celite is washed many times with 50-mL portions of technical-grade methylene chloride. In order to accelerate filtration and to improve washing, the surface of the Celite is disturbed from time to time with a spatula. The filtration time is approximately 50 min. The filtrate (300 mL) is then vigorously stirred for 1 hr in a mixture of 80 mL of 2 N sodium hydroxide and 40 mL of saturated aqueous sodium chloride. The organic phase [negative test for peroxides using a Merck kit (Merkoquant 10011)] is decanted, dried (magnesium sulfate or sodium sulfate), filtered and concentrated in a Büchi apparatus (bath temperature: 45°C) to leave 10 g of an oily material which is a mixture of methyl p-tolyl sulfoxide, 2-phenyl-2-propanol, and some starting sulfide. Optically pure sulfoxide is then easily isolated as follows:

The crude product (10 g) is diluted with 4 mL of a solvent mixture (ethyl acetate/cyclohexane = 9:1). This solution is poured onto a column (75-mm diameter) filled with 120 g of silica gel (Merck 230-400 mesh) for flash chromatography. Elution is performed under gravity and requires ~ 200 mL of the above solvent system, followed by ~ 200 mL of ethyl acetate. 2-Phenyl-2-propanol mixed with methyl p-tolyl sulfide is eluted in the first fraction (~ 150 mL, monitored by TLC). The subsequent fractions are collected (~ 300 mL)

50

and evaporated, giving 4.0 g of methyl p-tolyl sulfoxide $[\alpha]_D$ -88° (acetone, c 1), 89% ee, 85% yield. This material is crystallized once from 60 mL of hot hexane, to afford, after 2 hr at 20°C, 3.14 g (68%) of enantiomerically pure (Note 7) (S)-(-)-methyl p-tolyl sulfoxide as needles, $[\alpha]_D$ -142° (acetone, c 1), mp 73-76°C (Reichert microscope with heating system).

The chemical purity of the compound is checked by [1]NMR (250 MHz) and thin layer chromatography (silica gel, eluent: ethyl acetate) which show the complete absence of the corresponding sulfide and sulfone (Note 8).

2. Notes

1. Methylene chloride, technical grade (99.5%), was passed through a column of basic alumina (grade I) and then stored over molecular sieves (4 Å).

2. (-)-Diethyl (S,S)-tartrate was obtained from the Aldrich Chemical Company, Inc. and was distilled (bp 120°C/2 mm).

3. Titanium(IV) isopropoxide, Ti(OiPr)$_4$ (Aldrich Chemical Company, Inc.) was distilled under an inert atmosphere (nitrogen or argon) and stored in a flask with a septum cap under argon (bp 85°C/1.5 mm).

4. Methyl p-tolyl sulfide, available from Aldrich Chemical Company, Inc., was distilled (bp 95°C, 18 mm) before use.

5. Cumene hydroperoxide, obtained from Aldrich Chemical Company, Inc., technical grade (80%), was dried overnight over 4 Å molecular sieves (pellets) prior to use.

6. A standard freezer without accurate temperature control was used; it is estimated that the temperature is -23°C ± 1°C. One night is a convenient reaction time, but oxidation is in fact complete after a few hours.

7. Enantiomerically pure (R)-(+)-methyl p-tolyl sulfoxide, prepared from (-)-menthyl p-tolylsulfinate,[2-4] was described with the following specific rotations: $[\alpha]_D$ +145.5° (acetone),[3] $[\alpha]_D^{25}$ +168° (acetone, c 1.8),[4] $[\alpha]_D$ +189° (CHCl$_3$, c 1).[5] The submitters checked a sample prepared[3,4] and kindly provided by Professor G. Solladié (Strasbourg), which was recrystallized from hexane: $[\alpha]_D$ +146° ± 1 (acetone, c 1), mp 76-77°C. HPLC analysis carried out on a chiral stationary phase shows the absence of the enantiomer (Dr. Tambute, private communication). The same analysis shows that the product obtained by the procedure described above is of 99.5% ee.

8. The checkers attempted chromatography of the three-component mixture on a 150-mmol scale using 55- and 75-mm diameter columns. However, mixed fractions were obtained even with seemingly large R_f differences for the components.

3. Discussion

Both enantiomers of methyl p-tolyl sulfoxide are available from the above procedure by selection of the appropriate diethyl tartrate. This procedure describes the preparation of (S)-(-)-methyl p-tolyl sulfoxide which is not easy to prepare by the Andersen method[2-4] using (+)-menthol.

Cumene hydroperoxide was selected because it was recently observed[6] that it gives in many cases better ee's in asymmetric oxidation of sulfides than the original procedure with t-butyl hydroperoxide.[7-9]

The enantiomeric purity of the crude (S)-methyl p-tolyl sulfoxide produced from the oxidation is close to 90% (measurement made on a sample of 200 mg of material purified by flash chromatography on silica gel, eluent: ethyl acetate - ethanol = 96:4). However, when oxidation is performed on a

10-mmol scale, enantioselectivity is improved (96% ee, 87% isolated yield).[6] We have no explanation for this optimum scale effect. It could be because of easier temperature control on a small scale (there is a decrease of enantio-selectivity above -20°C or below -25°C).[7]

t-Butyl hydroperoxide (anhydrous toluene solution prepared as described in (9)) was also used as the oxidant on a 0.11-mol scale in the presence of 0.10 mol of methyl p-tolyl sulfide and 0.053 mol of chiral titanium complex (Ti/DET/H_2O = 1:2:1). The procedure is identical with the one described above using cumene hydroperoxide. The ee of the crude sulfoxide is 84%. Pure (S)-methyl p-tolyl sulfoxide $[\alpha]_D$ -146° (acetone, c 1) is obtained without flash chromatography by three recrystallizations of the crude material from hexane in 50% yield (with respect to sulfide).

Preparation of various enantiomerically pure sulfoxides by oxidation of sulfides seems feasible in the cases where asymmetric synthesis occurs with ee's in the range of 90% giving crystalline products which can usually be recrystallized up to 100% ee. Aryl methyl sulfides usually give excellent enantioselectivity during oxidation[6-9] and are good candidates for the present procedure. For example, we have shown on a 10-mmol scale that optically pure (S)-(-)-methyl phenyl sulfoxide $[\alpha]_D$ -146° (acetone, c 1)[6] could be obtained in 76% yield after oxidation with cumene hydroperoxide followed by flash chromatographic purification on silica gel and recrystallizations at low temperature in a mixed solvent (ether-pentane). Similarly (S)-(-)-methyl o-methoxyphenyl sulfoxide, $[\alpha]_D$ -339° (acetone, c 1.5 100% ee measured by HPLC), was obtained in 80% yield by recrystallizations from hexane.

The method with cumene hydroperoxide has been recently used with success,[10] to prepare both enantiomers of methyl p-methoxyphenyl sulfoxide which were then taken as starting material for the total synthesis of biological compounds.

53

1. Laboratoire de Synthèse Asymétrique, UA CNRS 255, Université Paris-Sud, 91405 Orsay, France.

2. Andersen, K. K. *Tetrahedron Lett.* **1962**, 93.

3. Mislow, K.; Green, M. M.; Laur, P.; Melillo, J. T.; Simmons, T.; Ternay, A. L., Jr. *J. Am. Chem. Soc.* **1965**, *87*, 1958.

4. Solladié, G. *Synthesis* **1981**, 185.

5. Tsuchihashi, G.-i.; Iriuchijima, S.; Ishibashi, M. *Tetrahedron Lett.* **1972**, 4605.

6. Zhao, S. H.; Samuel, O.; Kagan, H. B. *Tetrahedron* **1987**, *43*, 5135.

7. Pitchen, P.; Duñach, E.; Deshmukh, M. N.; Kagan, H. B. *J. Am. Chem. Soc.* **1984**, *106*, 8188.

8. Duñach, E.; Kagan, H. B. *Nouv. J. Chim.* **1985**, *9*, 1.

9. Kagan, H. B.; Duñach, E.; Nemecek, C.; Pitchen, P.; Samuel, O.; Zhao, S.-H. *Pure Appl. Chem.* **1985**, *57*, 1911.

10. Davis, R.; Kern, J. R.; Kurz, L. J.; Pfister, J. R.; *J. Am. Chem. Soc.* **1988**, *110*, 7873.

Appendix

Chemical Abstracts Nomenclature (Collective Index Number);
(Registry Number)

(S)-(-)-Methyl p-tolyl sulfoxide: Benzene, 1-methyl-4-(methylsulfinyl)-, (S)-
(9); (5056-07-5)

(S,S)-(-)-Diethyl tartrate: Tartaric acid, diethyl ester, (-)- or D- (8);
Butanedioic acid, 2,3-dihydroxy-, diethyl ester, [S-(R*,R*)]- (9);
(13811-71-7)

Titanium(IV) isopropoxide: Isopropyl alcohol, titanium(4+) salt (8);
2-Propanol, titanium(4+) salt (9); (546-68-9)

Methyl p-tolyl sulfide: Sulfide, methyl p-tolyl (8); Benzene, 1-methyl-4-
(methylthio)- (9); (623-13-2)

Cumene hydroperoxide: Hydroperoxide, α,α-dimethylbenzyl (8); hydroperoxide,
1-methyl-1-phenylethyl (9); (80-15-9)

YEAST REDUCTION OF 2,2-DIMETHYLCYCLOHEXANE-1,3-DIONE:

(S)-(+)-3-HYDROXY-2,2-DIMETHYLCYCLOHEXANONE

(Cyclohexanone, 3-hydroxy-2,2-dimethyl-, (S)-)

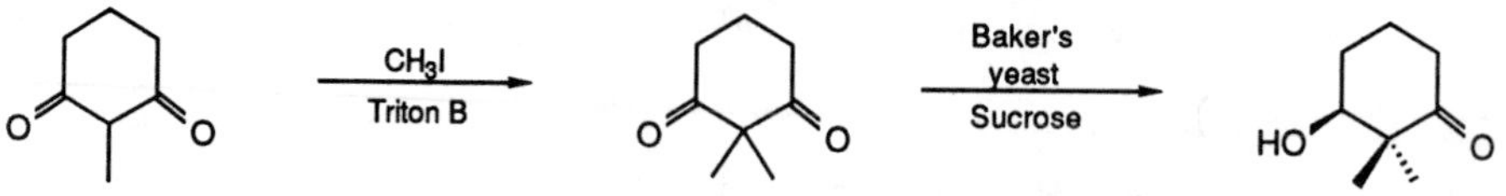

Submitted by Kenji Mori and Hideto Mori.[1]

Checked by Mark Hopkins and Larry E. Overman.

1. Procedure

A. 2,2-Dimethylcyclohexane-1,3-dione. In a 1-L, three-necked, round-bottomed flask equipped with a magnetic stirrer, 200-mL, pressure-equalizing dropping funnel and reflux condenser (the top of which is connected to a calcium chloride drying tube) are placed 50.4 g (0.4 mol) of 2-methyl-cyclohexane-1,3-dione[2] and 500 mL of dry methanol. To this stirred solution is added dropwise 168 mL of commercial Triton B (40% in methanol) (Notes 1 and 2). After the addition is complete, the resulting solution is stirred at room temperature for 10 min and 60.0 g (0.423 mol) of methyl iodide is added portionwise. This solution is then stirred and heated under reflux for 16-20 hr (Note 3). After the reaction mixture is cooled to room temperature, about 400 mL of methanol is removed by rotary evaporation. The residue is poured into a mixture of 100 mL (1.2 mol) of concd hydrochloric acid and about 100 g of ice to decompose the O-alkylated product (Note 4), and the mixture is stirred for 30 min. The precipitated solid (recovered starting material) is

56

collected by filtration and the filtrate is extracted four times with 100 mL of ethyl acetate. The combined ethyl acetate extracts are washed with 5% sodium thiosulfate solution (4 x 100 mL), saturated sodium hydrogen carbonate solution (2 x 100 mL) and saturated sodium chloride solution (100 mL), dried over anhydrous magnesium sulfate and filtered. Ethyl acetate is removed by rotary evaporation and the residue is distilled to give 30-32 g (54-57%) of 2,2-dimethylcyclohexane-1,3-dione, bp 92-97°C (4 mm) as a glassy solid (Notes 5 and 6).

B. *(S)-(+)-3-Hydroxy-2,2-dimethylcyclohexanone*. In a 5-L, three-necked, round-bottomed flask equipped with a mechanical stirrer, thermometer, and stopper are placed 3 L of tap water and 450 g of sucrose (Note 7). The mixture is stirred at 30°C, and 200 g of dry baker's yeast (Note 8) is added with stirring, whereupon brisk fermentation takes place (Note 2). This fermenting mixture is stirred at 30°C for 10 min and a solution of 15 g (0.107 mol) of 2,2-dimethylcyclohexane-1,3-dione in 95% ethanol (30 mL) and 0.2% Triton X-100 (120 mL) is added portionwise (Note 9). The mixture is stirred at 30°C for 40-48 hr (Note 10). Diethyl ether (about 200 mL) and Celite (about 50 g) are then added to the mixture and it is left to stand overnight. After the flocculated yeast cells are precipitated, the mixture is filtered through a pad of Celite (Note 11). The filter-cake is washed with ethyl acetate (2 x 100 mL) and the combined filtrate and washings are saturated with sodium chloride and extracted four times with 100 mL of ethyl acetate. The combined ethyl acetate extracts are washed with saturated sodium hydrogen carbonate solution (200 mL) and saturated sodium chloride solution (200 mL), dried over anhydrous magnesium sulfate and filtered. Ethyl acetate is removed by rotary evaporation and the residue (about 20 g) is chromatographed over 200 g of silica gel (Fuji Davison 820 MH gel) (Note 12).

Elution with hexane-ethyl acetate (10:1-5:1) gives 5-6 g of recovered starting material. Further elution with hexane-ethyl acetate (5:1-1:2) gives 7.7-7.9 g (47-52%) of (S)-3-hydroxy-2,2-dimethylcyclohexanone. An analytical sample can be obtained by distillation, bp 85-87°C/3.7 mm, $[\alpha]_D^{21}$ +23.0° (CHCl$_3$, c 2.0) (Notes 13, 14, and 15).

2. Notes

1.   The submitters used Triton B purchased from Tokyo Kasei Co., Ltd. whereas the checkers used material purchased from Aldrich Chemical Company, Inc.

2.   No temperature control is required because of the non-exothermic nature of this reaction.

3.   The checkers found (GLC analysis using a 12.5-m, 5% methyl silicone capillary column) that the reaction was complete within 4-6 hr.

4.   To ensure the complete decomposition of the O-alkylated product, this amount of hydrochloric acid is used.   A smaller amount may result in incomplete decomposition of the by-product.

5.   This material is sufficiently pure for use in the next step. Analysis by ^{13}C NMR indicates that 2-4% of an unknown impurity is present. Recrystallization from hexane-methylene chloride gives pure product melting at 37-38°C

6.   The spectra are as follows: ^{1}H NMR (250 MHz, CDCl$_3$) δ: 1.29 (s, 6 H), 1.93 (5 lines, 2 H, J = 6.5), 2.67 (t, 4 H, J = 6.9); ^{13}C NMR (76 MHz, CDCl$_3$) δ: 18.1, 22.3, 37.4, 61.8, 210.6.

7.   Commercially available sucrose is used.   The checkers employed deionized water.

8.   The submitters used yeast purchased from the Oriental Yeast Co., Ltd., whereas the checkers used Fleischman's Dry Active Yeast.

9.   The submitters used Triton X-100 purchased from Tokyo Kasei Co., Ltd. whereas the checkers used material purchased from Rohm & Haas Co.

10.   To keep the temperature at 30°C, the flask is gently heated on a large water bath.

11.   The checkers found that the use of coarse filter paper and a Buchner funnel was preferable to a sintered glass funnel.

12.   The checkers found that the chromatography was best accomplished using 450 g of silica gel (E. Merck).

13.   The spectral properties of (S)-(+)-3-hydroxy-2,2-dimethylcyclo-hexanone are as follows: IR ν_{max} (film) cm^{-1}: 3470 (s), 1705 (s), 1120 (m), 1055 (s), 985 (s), 965 (m); ^{1}H NMR (250 MHz, $CDCl_3$) δ: 1.11 (s, 3 H), 1.15 (s, 3 H), 1.60-1.71 (m, 1 H), 1.76-1.86 (m, 1 H), 1.96-2.05 (m, 2 H), 2.16 (br s, 1 H), 2.35-2.45 (m, 2 H), 3.69 (dd, 1 H, J = 7.6, 2.9); ^{13}C NMR (76 MHz, $CDCl_3$) δ: 19.7, 20.7, 22.9, 29.0, 37.3, 51.3, 77.8, 215.3.

14.   The optical purity of (S)-(+)-3-hydroxy-2,2-dimethylcyclohexanone can be determined by HPLC analysis.   The (S)-α-methoxy-α-trifluoromethyl-phenylacetate (MTPA ester) is prepared according to the reported procedure:[3] HPLC analysis (Column, Nucleosil® 50-5, 25 cm x 4.6 mm; eluent, hexane:THF = 30:1, 1.03 mL/min; detected at UV 256 nm) retention time 25.6 min (98.0-99.4%) and 29.6 min (0.6-2.0%).   Therefore the optical purity is determined to be 96.0-98.8% ee.

15. Analysis of the MTPA ester of this product by 250 MHz ^{1}H NMR and capillary GLC (12.5 m, 5% methyl silicone column) failed to detect any more of the minor diastereomer than would have been expected from the purity (98% ee) of the MTPA-Cl employed.

3. Discussion

Microbial reduction of prochiral cyclopentane- and cyclohexane-1,3-diones was extensively studied during the 1960's in connection with steroid total synthesis.[4] Kieslich, Djerassi, and their coworkers reported the reduction of 2,2-dimethylcyclohexane-1,3-dione with *Kloeckera magna* ATCC 20109, and obtained (S)-3-hydroxy-2,2-dimethylcyclohexanone.[5] We found that the reduction of the 1,3-diketone can also be effected with conventional baker's yeast, and secured the hydroxy ketone of 98-99% ee as determined by an HPLC analysis of the corresponding (S)-α-methoxy-α-trifluoromethylphenylacetate (MTPA ester).[6,7] (S)-3-Hydroxy-2,2-dimethylcyclohexanone has been proved to be a versatile chiral non-racemic building block in terpene synthesis as shown in Figure 1.

Figure 1. Natural Products Synthesized from (S)-3-Hydroxy-2,2-dimethylcyclohexanone

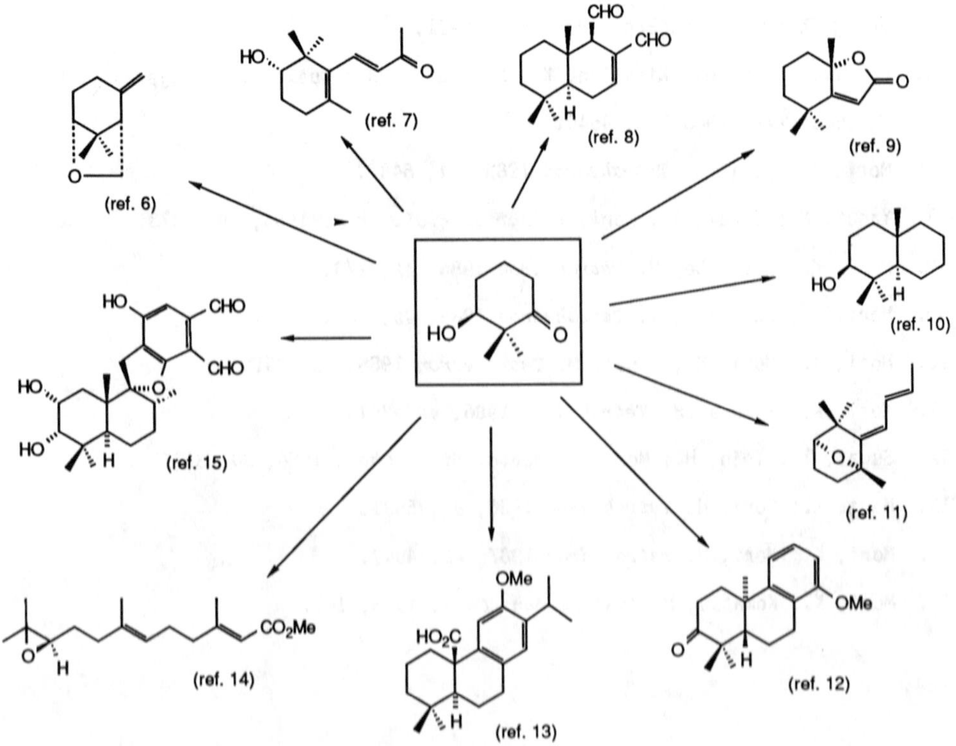

1. Department of Agricultural Chemistry, The University of Tokyo, Yayoi 1-1-1, Bunkyo-Ku, Tokyo 113, Japan.

2. Mekler, A. B.; Ramachandran, S.; Swaminathan, S.; Newman, M. S. *Org. Synth., Collect. Vol. 5*, **1973**, 743.

3. Dale, J. A.; Mosher, H. S. *J. Am. Chem. Soc.* **1973**, *95*, 512.

4. Kieslich, K. "Microbial Transformations of Non-Steroid Cyclic Compounds;" Georg Thieme: Stuttgart, 1976; pp. 28-31.

5. Lu, Y.; Barth, G.; Kieslich, K.; Strong, P. D.; Duax, W. L.; Djerassi, C. *J. Org. Chem.* **1983**, *48*, 4549.

6. Mori, K.; Mori, H. *Tetrahedron* **1985**, *41*, 5487.

7. Yanai, M.; Sugai, T.; Mori, K. *Agric. Biol. Chem.* **1985**, *49*, 2373.

8. Mori, K.; Watanabe, H. *Tetrahedron* **1986**, *42*, 273.

9. Mori, K.; Nakazono, Y. *Tetrahedron* **1986**, *42*, 283.

10. Mori, K.; Mori, H.; Yanai, M. *Tetrahedron* **1986**, *42*, 291.

11. Mori, K.; Tamura, H. *Tetrahedron* **1986**, *42*, 2643.

12. Sugai, T.; Tojo, H.; Mori, K. *Agric. Biol. Chem.* **1986**, *50*, 3127.

13. Mori, K.; Mori, H. *Tetrahedron* **1986**, *42*, 5531.

14. Mori, K.; Mori, H. *Tetrahedron* **1987**, *43*, 4097.

15. Mori, K.; Komatsu, M. *Liebigs Ann. Chem.* **1988**, 107.

Appendix

Chemical Abstracts Nomenclature (Collective Index Number); (Registry Number)

(S)-(+)-3-Hydroxy-2,2-dimethylcyclohexanone: Cyclohexanone, 3-hydroxy-2,2-dimethyl-, (S)- (11); (87655-21-8)

2,2-Dimethylcyclohexane-1,3-dione: 1,3-Cyclohexanedione, 2,2-dimethyl- (8,9); (562-13-0)

2-Methylcyclohexane-1,3-dione: 1,3-Cyclohexanedione, 2-methyl- (8,9); (1193-55-1)

Triton B: Ammonium, benzyltrimethyl-, hydroxide (8); Benzenemethanaminium, N,N,N-trimethyl-, hydroxide (9); (100-85-6)

Sucrose (8); α-D-Glucopyranoside, β-D-fructofuranosyl (9); (57-50-1)

Triton X-100: Glycols, polyethylene, mono[p-(1,1,3,3-tetramethylbutyl)-phenyl]ether (8); Poly(oxy-1,2-ethanediyl), α-[4-(1,1,3,3-tetramethylbutyl)-phenyl]-ω-hydroxy- (9); (9002-93-1)

DIRECTED HOMOGENEOUS HYDROGENATION: METHYL
anti-3-HYDROXY-2-METHYLPENTANOATE
(Pentanoic acid, 3-hydroxy-2-methyl-, methyl ester, (R*,R*)-(±)-)

A.

B.

C.

Submitted by John M. Brown, Phillip L. Evans, and Alun P. James.[1]

Checked by Ulrike Eggert, H. M. R. Hoffmann, and Ekkehard Winterfeldt.

1. Procedure

A.   Bicyclo[2.2.1]hepta-2,5-diene   1,4-bis(diphenylphosphino)butane-
rhodium trifluoromethanesulfonate, 1.   A 250-mL, two-necked, round-bottomed
flask equipped with a septum, two-way stopcock used as a gas inlet and outlet,
and a magnetic stirrer is charged with a solution of bicyclo[2.2.1]hepta-2,5-
diene-2,4-pentanedionatorhodium (117.6 mg, 0.4 mmol) (Note 1) in dry
tetrahydrofuran (2.5 mL) under a gentle stream of argon.   Trimethylsilyl

64

trifluoromethanesulfonate (97.8 mg, 0.44 mmol) (Note 2) is added in one portion by microsyringe via the septum, resulting in a color change from yellow to orange. Solid 1,4-bis(diphenylphosphino)butane (170.4 mg, 0.4 mmol) (Note 3) is added all at once, with removal and immediate replacement of the septum cap. The color of the solution darkens to a deep orange-red, and precipitation of an orange solid occurs over 1-2 min. Dry 30-40 petroleum ether (10 mL) is added with vigorous stirring. The mixture is allowed to settle and the solvent is removed first by syringe and finally under reduced pressure with an oil pump. Argon is admitted and the catalyst is dried and stored under argon (Notes 4, 5).

B. *Methyl 3-hydroxy-2-methylenepentanoate.* A 250-mL, round-bottomed flask is charged with methyl acrylate (50.0 mL, 0.556 mol), propionaldehyde (60.0 mL, 0.832 mol), and 1,4-diazabicyclo[2.2.2]octane (3.0 g, 26.8 mmol) (Note 6). After the solution is stirred briefly to ensure complete dissolution, the flask is stoppered and set aside at ambient temperature for 7 days. The reaction mixture is then dissolved in dichloromethane (150 mL), washed with 1 M hydrochloric acid (100 mL), and the organic layer is separated and dried over anhydrous magnesium sulfate. Solvent is removed under reduced pressure after filtration and the residue is fractionated through a 12"-vacuum-jacketed Vigreux column. The main fraction boiling at 54°C (0.1 mm) is collected as a water-white liquid to afford 57 g (71%) of condensation product (Note 7). Care must be taken in the distillation to avoid contamination by high-boiling impurities which may inactivate the catalyst.

C. *Methyl dl-anti-3-hydroxy-2-methylpentanoate.* Freshly distilled methyl 3-hydroxy-2-methylenepentanoate (14.4 g, 0.1 mol) is added to the biphosphinorhodium catalyst (catalyst/substrate ratio 1/250) (Note 8) in the flask from step A. Methanol (40 mL, distilled from $Mg(OMe)_2$) is added. The

65

apparatus is sealed with a rubber septum, the side-arm is connected to a burette line, and the apparatus is then transferred to a dry ice/2-propanol bath. The vessel is evacuated to 1 mm three times and filled with hydrogen (Note 9) each time, the burette being partially filled on the last occasion. The mixture is warmed to near ambient temperature and transferred to a water bath at 12 ± 2°C (Note 10). When thermal equilibration is complete, the stirrer is started so as to create a deep vortex in the reaction solution, which darkens to a brick-red color with rapid uptake of hydrogen. The burette is recharged with hydrogen as necessary, and gas absorption occurs steadily until the reaction is complete, when ca. 2.5 L of hydrogen has been consumed (Note 11). Completion of reaction can readily be checked by addition of 50 μL of methyl 3-hydroxy-2-methylenepentanoate to the reaction solution by microsyringe, leading to a perceptible burst of gas absorption. At this stage the flask is disconnected from the hydrogen line, flushed with argon and the contents are transferred to a 250-mL, round-bottomed flask. Solvent is removed at ambient temperature on a rotary evaporator, and the residue is dissolved in diethyl ether (40 mL) and 30-40 petroleum ether (150 mL). The mixture is filtered through a small plug of silica (60 μ, Merck flash chromatography grade) which effectively retains the residual rhodium catalyst (Note 1). Solvents are removed on a rotary evaporator, and the colorless product (Note 12) is distilled in a Kugelrohr apparatus, bp 50°C (~ 0.5 mm). The yield of methyl dl-anti-3-hydroxy-2-methylpentanoate is 13.3 g (91%), chemically and stereochemically pure by ^{13}C NMR (Note 13).

This procedure may be adapted for kinetic resolution of the reactant, employing an optically active biphosphinerhodium catalyst (Note 13).

2. Notes

1.   The starting material for complex preparation is $RhCl_3 \cdot 3\ H_2O$, obtained as a loan from Johnson Matthey Co.   Rhodium-containing reaction residues are collected for return. The synthesis of bicyclo[2.2.1]hepta-2,5-diene-2,4-pentanedionatorhodium has been described by Wilkinson,[2] and later by Green,[2] and may be carried out by the following modification: $RhCl_3 \cdot 3\ H_2O$ (0.68 g, 2.58 mmol) was dissolved in 90% ethanol (10 mL) and stirred with freshly distilled bicyclo[2.2.1]hepta-2,5-diene (1.95 mL) for 2 days under argon.   The yellow precipitate was filtered, dried and dissolved in tetrahydrofuran (10 mL) to which was added sodium 2,4-pentanedionate (0.314 g, 2.58 mmol) in one portion. The suspension was stirred vigorously for 4 hr, filtered and solvent removed from the filtrate under reduced pressure. The resulting complex [530 mg, 70%, mp 172-175°C (lit.[2] 175-177°C)] is sufficiently pure to use, but may be sublimed under reduced pressure if desired.   Alternatively, catalysts may be purchased from Chemical Products, Johnson Matthey Co., Royston, Cambridgeshire, England.

2.   This toxic and corrosive reagent (Aldrich Chemical Company, Inc.) was transferred directly from the septum-sealed commercial sample.

3.   This reagent was obtained from the Strem Chemical Company, Inc.

4.   If it is desired to isolate and store the catalyst, an alternative procedure, preferred by the submitters, may be used. A 25-mL Schlenk tube is used as a reaction flask.   After addition of the petroleum ether, the resulting orange suspension is filtered by centrifugation in a Craig tube and dried under argon. The bright orange solid (0.26-0.28 g), mp 211-212°C (dec), is indefinitely stable when stored in a -20°C freezer under argon.

5. Earlier work on directed hydrogenation used tetrafluoroborate salts.[3] Triflate salts seem superior in keeping properties, and their preparation is easy and convenient.

6. Methyl acrylate, 99% (stabilized with 200 ppm hydroquinone monomethyl ether), propionaldehyde 97%, and 1,4-diazabicyclo[2.2.2]octane were purchased from the Aldrich Chemical Company, Inc. and used as supplied.

7. The Michael-induced condensation reaction between acrylates and aldehydes[4] is dramatically accelerated by high pressure, cutting the reaction time from several days to a few minutes.[5]

8. The submitters used a 1:500 catalyst/substrate ratio, (H_2 uptake ca. 40 mL/min) but the checkers used the higher ratio to speed up the hydrogen uptake. If an accurate rate of hydrogen uptake is crucial to a user, the lower ratio may be preferred.

9. Hydrogen of 99.99% purity, supplied by the British Oxygen Company, was employed.

10. The diastereoselectivity of reduction increases with decreasing temperature, and the conditions chosen represent a compromise between rate and specificity.

11. On one occasion the submitters added an extra 15.0 g of starting material to the reaction vessel at this point. Hydrogenation proceeded to completion (i.e., 1000 turnovers, in total) but slowed appreciably in the latter stages of reaction.

12. The submitters fractionated the product through a short Vigreux column, collecting the main fraction boiling at 65°C (1 mm). The ^{13}C NMR spectrum of the bulk reaction product is as follows: ($CDCl_3$) δ: 8.89 (CH_3CH), 13.19 (CH_3CH_2), 26.41 (CH_2), 44.21 (OCH_3), 50.79 (CH-CO), 73.73 (CH-O), 175.63 (C=O). The syn isomer exhibits resonances at ($CDCl_3$) δ: 9.10, 13.19, 28.33,

68

44.21, 50.94, 71.53; it may be prepared by Pd/C/H$_2$ reduction of the starting material and separation of the diastereomers of product by preparative GLC (OV 225, 15', 150°C).

13. When (R,R)-1,2-bis(o-anisylphenylphosphino)ethanerhodium triflate (DIPAMPRh[+]) was used as catalyst,[6] 5.0 g of starting material was hydrogenated with 0.1 g of catalyst in methanol at 0°C to 65% reaction (ca. 6 hr). Workup and isolation by preparative GLC (OV 225, 15', 150°C) gave 2.0 g of 2R,3R-(-)-methyl 3-hydroxy-2-methylpentanoate, $[\alpha]_{587}^{20}$ -6.4° (chloroform, c 4) and 0.8 g of recovered S-(-)-methyl 3-hydroxy-2-methylenepentanoate, $[\alpha]_{587}^{20}$ -20.3° (chloroform, c 2). The optical purity of the former is 57 ± 2% by chiral shift NMR (Eu(hfc)$_3$) and the latter is $\geqslant$ 97% optically pure. This represents an enantiomer rate ratio of 13:1.

3. Discussion

α-Substituted β-hydroxy esters are the formal product of an ester enolate aldol condensation. High anti-stereoselectivity in the reaction requires the lithium enolate of a reasonably bulky aryl ester and a sterically demanding aldehyde. The condensation between 2,6-dimethylphenyl propionate and 2-methylpropanal has been described in *Organic Syntheses*,[7] under conditions where the product is formed with 98% anti-selectivity. Recently the condensation of E-silylketene acetals derived from N-methylephedrine esters with aldehydes mediated by titanium(IV) chloride has been shown to occur with good anti-selectivity and in high ee.[8]

Alternatively, the anti-α-alkyl-β-hydroxy ester structure may be obtained by alkylation of the dianion of a β-hydroxy ester, which occurs with $\geqslant$ 95% stereoselectivity.[9] Since the starting materials are available in moderate to high optical purity by yeast reduction of β-keto esters, [10a,b] this constitutes an asymmetric synthesis.[10c]

The present procedure involving homogeneous catalysis is operationally simple and takes advantage of the easy availability of 2-(1'-hydroxyalkyl)-acrylic esters. A two-step procedure involving kinetic resolution of the racemic starting material with an optically active hydrogenation catalyst, followed by a further reduction with an achiral catalyst, leads to diastereomerically pure products in $\geq$ 97% ee.

Directed hydrogenation is applicable to olefins which are α'-disubstituted or trisubstituted with a polar functional group at an adjacent chiral center.[11] The latter must be capable of sustained coordination to the metal during the catalytic cycle, thereby exerting stereochemical control on hydride transfer to carbon. The presence of an electron-withdrawing group at the α'-position enhances both the reaction rate and stereoselectivity, making 2-(1'-hydroxyalkyl)acrylates highly suitable substrates for the reaction. The polar group is most commonly -OH, but CO_2Me, $CONR_2$, NHCOR and $NHCO_2R$ have all been used. Other substituents including OMe, OCOR and NH_2 are much less effective. For acyclic cases where the reactant possesses conformational flexibility, cationic rhodium complexes derived from the 7-ring chelate of 1,4-bis(diphenylphosphino)butane have proved most effective. With cyclic reactants cationic iridium catalysts of the type introduced by Crabtree and Morris[12] have generally been more successful, and the procedure is more tolerant of steric bulk in the reactant olefin. A series of examples is collected in Table I.

70

Substituted acrylates (which resemble the enamide substrates employed in asymmetric hydrogenation)[13] may be deracemized by reduction with an optically active catalyst, especially DIPAMPRh$^+$. Selectivity ratios of 12:1 to 22:1 have been obtained for a variety of reactants; with compounds of reasonable volatility, separation of starting material and product may be effected by preparative GLC. Recovered starting material can then be reduced with an achiral catalyst to give the optically pure anti product. Examples of kinetic resolutions by this method are given in Table II. More recently very successful kinetic resolutions of allylic alcohols have been carried out with Ru(BINAP) catalysts.[10c]

TABLE I

Directed Homogeneous Hydrogenation

Reactant	Product	Catalyst[a] (mol %)	Selectivity
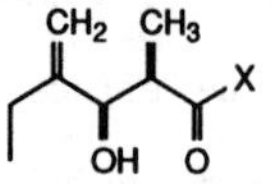 CH₂ CH₃ ... X ... OH O	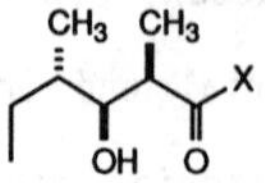 CH₃ CH₃ ... X ... OH O	A (>2)	93[3b]
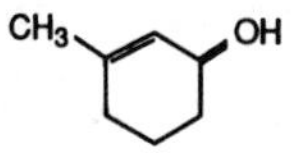 CH₃ ... OH	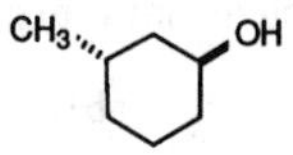 CH₃ ... OH	A (20)	99.6[3b]
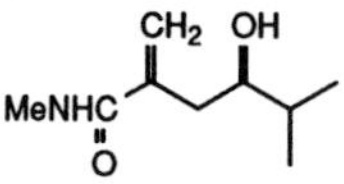 CH₂ OH ... MeNHC ... O	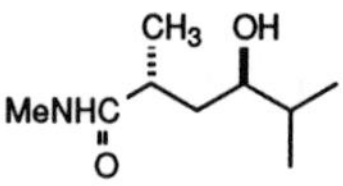 CH₃ OH ... MeNHC ... O	A (2)	92[13]
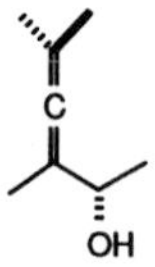 ... OH	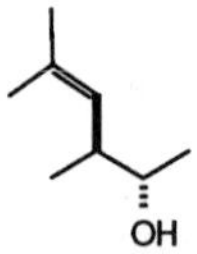 ... OH	A (3)	97[13]
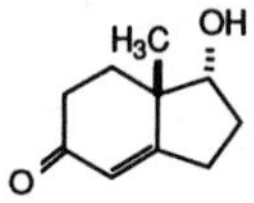 H₃C OH ... O	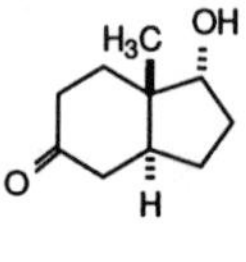 H₃C OH ... O ... H	B (20)	96[14]
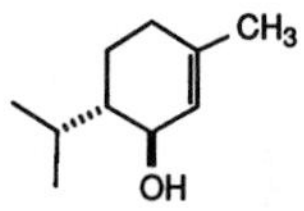 CH₃ ... OH	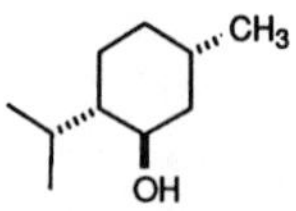 CH₃ ... OH	B (2.5)	99.7[15]
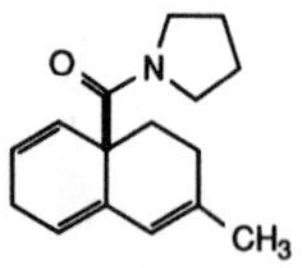 O N ... CH₃	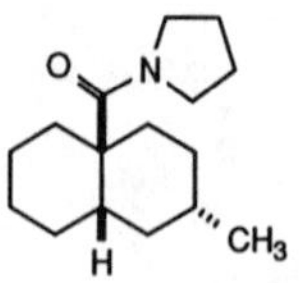 O N ... H ... CH₃	B (5)	99.9[16]

[a]Catalyst A is complex **1** (see text); catalyst B is (PCx₃)(py)(C₈H₁₂)IrPF₆.

72

TABLE II

Kinetic Resolutions in Acrylate Hydrogenation[a]

Recovered reactant	Major product	% Reaction	Ee
CH_2 / MeO_2C ... OH	CH_3 / MeO_2C ... OH	65	98[13]
CH_2 / MeO_2C ... CO_2Me	CH_3 / MeO_2C ... CO_2Me	65	96[17]
CH_2 / MeO_2C ... $NHAc$	CH_3 / MeO_2C ... $NHAc$	58	96[18]

[a] 1 - 4 mol % DIPAMPRh[+] in MeOH, usually at $0^\circ C$.

1. The Dyson Perrins Laboratory, University of Oxford, South Parks Road, Oxford OX1, 3QY, United Kingdom.

2. Bonati, F.; Wilkinson, G. *J. Chem. Soc.* **1964**, 3156; Green M.; Kuc, T. A.; Taylor, S. H. *J. Chem. Soc. (A)* **1971**, 2334.

3. (a) Brown, J. M.; Naik, R. G. *J. Chem. Soc., Chem. Commun.* **1982**, 348; (b) Evans, D. A.; Morrissey, M. M. *J. Am. Chem. Soc.* **1984**, *106*, 3866.

4. Drewes, S. E.; Emslie, N. D. *J. Chem. Soc., Perkin Trans I* **1982**, 2079; Baylis, A. B.; Hillman, M. E. German Patent, 2155/13; *Chem. Abstr.* **1972**, *77*, 34174q; Hoffmann, H. M. R.; Rabe, J. *J. Org. Chem.* **1985**, *50*, 3849.

5. Hill, J. S.; Isaacs, N. S. *Tetrahedron Lett.* **1986**, *27*, 5007.

6. Brown, J. M.; Cutting, I.; Evans, P. L.; Maddox, P. J. *Tetrahedron Lett.* **1986**, *27*, 3307.

7. Montgomery, S. H.; Pirrung, M. C.; Heathcock, C. H. *Org. Synth.* **1985**, *63*, 99; Heathcock, C. H.; Pirrung, M. C.; Montgomery, S. H.; Lampe, J. *Tetrahedron* **1981**, *37*, 4087.

8. Gennari, C.; Bernardi, A.; Colombo, L.; Scolastico, C. *J. Am. Chem. Soc.* **1985**, *107*, 5812.

9. Seebach, D.; Aebi, J.; Wasmuth, D. *Org. Synth.* **1985**, *63*, 109; Züger, M.; Weller, T.; Seebach, D. *Helv. Chim. Acta* **1980**, *63*, 2005; Wasmuth, D.; Arigoni, D.; Seebach, D. *Helv. Chim. Acta* **1982**, *65*, 344.

10. a) Seebach, D.; Sutter, M. A.; Weber, R. H.; Züger, M. F. *Org. Synth.* **1985**, *63*, 1; b) Wipf, B.; Kupfer, E.; Bertazzi, R.; Leuenberger, H. G. W. *Helv. Chim. Acta* **1983**, *66*, 485; c) For an alternative hydrogenation procedure see Kitamara, M.; Kasahara, I.; Manabe, K.; Noyori, R.; Takaya, H. *J. Org. Chem.* **1988**, *53*, 708.

11. Brown, J. M. *Angew. Chem., Intern. Ed. Engl.* **1987**, *26*, 190.

74

12. Crabtree, R. H; Felkin, H.; Morris, G. E. *J. Organomet. Chem.* **1977**, *141*, 205; Crabtree, R. H. *Acc. Chem. Res.* **1979**, *12*, 331.

13. Brown, J. M.; James, A. P.; Wali, M., to be published; Birtwistle, D. H.; Brown, J. M.; Herbert, R. H.; James, A. P.; Lee, K. F.; Taylor, R. J. *J. Chem. Soc. Chem. Commun.* **1989**, in press.

14. Stork, G.; Kahne, D. E. *J. Am. Chem. Soc.* **1983**, *105*, 1072.

15. Crabtree, R. H.; Davis, M. W. *J. Org. Chem.* **1986**, *51*, 2655.

16. Schultz, A. G.; McCloskey, P. J. *J. Org. Chem.* **1985**, *50*, 5905.

17. Brown, J. M.; James, A. P. *J. Chem. Soc., Chem. Commun.* **1987**, 181.

18. Brown, J. M.; James, A. P.; Prior, L. M. *Tetrahedron Lett.* **1987**, *28*, 2179.

Appendix

Chemical Abstracts Nomenclature (Collective Index Number)

(Registry Number)

Methyl anti-3-hydroxy-2-methylpentanoate: Pentanoic acid, 3-hydroxy-2-methyl, methyl ester, (R*,R*)-(±)- (11); (100992-75-4)

Bicyclo[2.2.1]hepta-2,5-diene-1,4-bis(diphenylphosphino)butanerhodium trifluoromethanesulfonate: Rhodium(1+), [(2,3,5,6-)-bicyclo[2.2.1]hepta-2,5-diene][1,4-butanediylbis[diphenylphosphine]-P,P']-, trifluoromethanesulfonate

Bicyclo[2.2.1]hepta-2,5-diene-2,4-pentanedionatorhodium: Rhodium, (2,5-norbornadiene) (2,4-pentanedionato)- (8); Rhodium, [(2,3,5,6-)-bicyclo[2.2.1]hepta-2,5-diene] (2,4-pentanedionato-0,0')- (9); (32354-50-0)

Trimethylsilyl trifluoromethanesulfonate: Methanesulfonic acid, trifluoro-, trimethylsilyl ester (8,9); (27607-77-8)

1,4-Bis(diphenylphosphino)butane: Phosphine, tetramethylenebis[diphenyl- (8); Phosphine, 1,4-butanediylbis[diphenyl- (9); (7688-25-7)

Methyl 3-hydroxy-2-methylenepentanoate: Pentanoic acid, 3-hydroxy-2-methylene-, methyl ester (9); (18052-21-6)

1,4-Diazabicyclo[2.2.2]octane (8,9); (280-57-9)

(S)-4-(PHENYLMETHYL)-2-OXAZOLIDINONE

(2-Oxazolidinone, 4-(phenylmethyl)-, (S)-)

A.

$$BF_3 \cdot OEt_2 \quad / \quad BH_3 \cdot SMe_2$$

B.

$$(EtO)_2CO \quad / \quad K_2CO_3$$

Submitted by James R. Gage and David A. Evans.[1]

Checked by Philip G. Meister and Leo A. Paquette.

1. Procedure

Caution! This reaction should be carried out in a hood since dimethyl sulfide is liberated during the course of the reaction.

A. *(S)-Phenylalanol.* A dry, 3-L, three-necked flask is equipped with a mechanical stirrer and a reflux condenser connected to a mineral oil bubbler. The flask is loaded with 165 g (1.00 mol) of (S)-phenylalanine (Note 1), then equipped with a 250-mL pressure-equalized addition funnel capped with a rubber septum through which is inserted a nitrogen-inlet needle. The flask is swept with nitrogen and filled with 500 mL of anhydrous tetrahydrofuran, and the addition funnel is charged with 123 mL (1.00 mol) of freshly distilled boron trifluoride etherate via cannula (Note 2). The boron trifluoride etherate is added dropwise to the phenylalanine slurry over a 30-min period with stirring, and the mixture is heated at reflux for 2 hr, resulting in a

77

colorless, homogeneous solution. The addition funnel is then charged via cannula with 88 g (110 mL, 1.15 mol) of 10 M borane-dimethyl sulfide complex (Note 3), which is added carefully to the *vigorously* refluxing solution over a 100-min period. During the course of the addition there is continuous evolution of dimethyl sulfide and hydrogen gas, and the solution turns from orange to light brown. A vigorous exotherm occurs approximately halfway through the addition period. (*Caution!* Note 4). The solution is heated at reflux for an additional 6 hr after the addition is complete (Notes 5 and 6), then allowed to cool to ambient temperature. The excess borane is quenched by the slow addition of 125 mL of a 1:1 tetrahydrofuran-water solution followed by 750 mL of 5 M aqueous sodium hydroxide. The resulting two-phase mixture is heated at reflux for 12 hr, cooled to room temperature, and filtered through a coarse fritted funnel. The residual solids are washed with two 25-mL portions of tetrahydrofuran, and the filtrate is concentrated on a rotary evaporator to remove the bulk of the tetrahydrofuran. The resulting slurry is extracted with one 400-mL and three 200-mL portions of dichloromethane. The combined organic extracts are dried over anhydrous sodium sulfate, filtered, and concentrated by rotary evaporation, yielding 141-158 g (93-104%) of a white crystalline solid which is recrystallized from ca. 600 mL of ethyl acetate to give 111-113 g (73-75%) of the desired product as white needles in two crops, mp 88.5-91°C (Note 7).

B. *(S)-4-(Phenylmethyl)-2-oxazolidinone.* A dry, 1-L, three-necked flask is equipped with a mechanical stirrer and a 12-in Vigreux column fitted with a distillation head and 200-mL receiver flask connected to a nitrogen source and bubbler. The flask is charged with 151 g (1.00 mol) of (S)-phenylalanol, 13.8 g (0.100 mol) of anhydrous potassium carbonate, and 250 mL (2.06 mol) of diethyl carbonate (Note 8). The mixture is lowered into an oil bath,

78

preheated to 135°C, and is stirred until dissolution is achieved (ca. 5 min). The distillation receiver is cooled in an ice bath, and ca. 120 mL of ethanol is collected from the reaction over a 2.5-hr period. The oil bath is removed upon cessation of the ethanol distillation. After the light yellow solution is cooled to ambient temperature, it is diluted with 750 mL of dichloromethane, transferred to a separatory funnel, and washed with 750 mL of water. The organic phase is dried over anhydrous magnesium sulfate, filtered, and concentrated on the rotary evaporator affording 200 g (113%) of a white crystalline solid. This material is taken up into 600 mL of a hot 2:1 ethyl acetate-hexane solution, filtered while hot, then allowed to crystallize to afford 136-138 g (78-79%) of large white plates, mp 84.5-86.5°C (Notes 9 and 10).

2. Notes

1.   (S)-Phenylalanine was obtained by the submitters from Ajinomoto Company, Inc.   The starting material obtained by the checkers from Sigma Chemical Company was dried under reduced pressure over phosphorus pentoxide for 2 days.

2.   Reagent grade tetrahydrofuran (Fisher Scientific Company) was either freshly distilled from sodium metal and benzophenone or dried for at least 24 hr over activated Linde 4 Å molecular sieves.   Boron trifluoride etherate was redistilled prior to use.   Fresh bottles of redistilled boron trifluoride etherate purchased from Aldrich Chemical Company, Inc., also usually give good results.

3.   Borane-dimethyl sulfide (10 M) was purchased from Aldrich Chemical Company, Inc. and used as received.

4. The potential vigor of this exotherm cannot be overemphasized. It occurs later and is correspondingly stronger if vigorous reflux is not maintained during the addition. The reaction mixture should be watched closely throughout the addition of borane, and addition should be temporarily suspended at the onset of the exotherm.

5. Dimethyl sulfide can be collected if desired by inserting a trap cooled in an acetone-dry ice bath into the hose leading to the bubbler.

6. Yields sometimes drop when an old bottle of borane-dimethyl sulfide is used. Reaction progress can be monitored by thin layer chromatography (silica gel, eluting with 10:10:1 chloroform-methanol-concentrated ammonium hydroxide). Any remaining phenylalanine stains heavily when exposed to ninhydrin (R_f = 0.35). If phenylalanine is detected after 5 hr of reflux, an additional 10 mL (0.10 mol) of borane-dimethyl sulfide is added via syringe, and the solution is heated at reflux for 1 additional hr.

7. The product has the following spectroscopic properties: IR (solution in dichloromethane) cm^{-1}: 3625, 3360, 3035, 2930, 2855, 1497, 1456, 1032; ^{1}H NMR (CDCl$_3$) δ: 1.5-2.0 (broad s, 3 H, NH$_2$, OH), 2.5 (dd, 1 H, HCHC$_6$H$_5$), 2.8 (dd, 1 H, HCHC$_6$H$_5$), 3.1 (m, 1 H, CHNH$_2$), 3.4 (dd, 1 H, HCHOH), 3.7 (dd, 1 H, HCHOH), 7.1-7.4 (m, 5 H, ArH): $[\alpha]_D$ -24.7° (ethanol, c 1.03). The checkers recorded $[\alpha]_D$ -22.4° (ethanol, c 1.03).

8. Diethyl carbonate (99%) was used as received from Aldrich Chemical Company, Inc.

9. The product has the following spectroscopic properties: IR (solution in dichloromethane) cm^{-1}: 3460, 3020, 1760, 1480, 1405, 1220; ^{1}H NMR (CDCl$_3$) δ: 2.9 (d, 2 H, CH$_2$C$_6$H$_5$), 4.0-4.6 (m, 3 H, CHCH$_2$O), 5.6 (broad s, 1 H, NH), 7.1-7.5 (m, 5 H, ArH); $[\alpha]_D$ +4.9° (ethanol, c 1.10).

80

10. The enantiomeric excess was determined to be >99% by capillary GLC analysis (30 m x 32 mm WCOT column coated with Carbowax 20 M, hydrogen carrier gas, linear velocity ca. 94 cm/s, oven temperature 225°C) of the imide derived from the Mosher acid chloride.[2]

3. Discussion

The utilization of α-amino acids and their derived β-amino alcohols in asymmetric synthesis has been extensive.[3] A number of procedures have been reported for the reduction of a variety of amino acid derivatives; however, the direct reduction of α-amino acids with borane has proven to be exceptionally convenient for laboratory-scale reactions.[4] These reductions characteristically proceed in high yield with no perceptible racemization. The resulting β-amino alcohols can, in turn, be transformed into oxazolidinones, which have proven to be versatile chiral auxiliaries. Besides the highly diastereoselective aldol addition reactions,[5] enolates of N-acyl oxazolidinones have been used in conjunction with asymmetric alkylations,[6] halogenations,[7] hydroxylations,[8] acylations,[9] and azide transfer processes,[10] all of which proceed with excellent levels of stereoselectivity.

The phenylalanine-derived oxazolidinone featured here enjoys three practical advantages over the valine-derived oxazolidinone developed earlier in this laboratory.[5] First, both the intermediate β-amino alcohol and the derived oxazolidinone are crystalline solids which can be purified conveniently by direct crystallization. Second, the oxazolidinone contains a UV chromophore which greatly facilitates TLC or HPLC analysis when it is employed as a chiral auxiliary. Finally, both enantiomers of phenylalanine are readily available, enabling stereocontrol in either sense simply by using the oxazolidinone derived from the appropriate enantiomer.

1. Department of Chemistry, Harvard University, Cambridge, MA 02138.

2. Dale, J. A.; Dull, D. L.; Mosher, H. S. *J. Org. Chem.* **1969**, *34*, 2543.

3. Coppola, G. M.; Schuster, H. F. "Asymmetric Synthesis: Construction of Chiral Molecules Using Amino Acids;" Wiley: New York, 1987.

4. Smith, G. A.; Gawley, R. E. *Org. Synth.* **1985**, *63*, 136.

5. Evans, D. A.; Bartroli, J.; Shih, T. L. *J. Am. Chem. Soc.* **1981**, *103*, 2127.

6. Evans, D. A.; Ennis, M. D.; Mathre, D. J. *J. Am. Chem. Soc.* **1982**, *104*, 1737.

7. Evans, D. A.; Ellman, J. A.; Dorow, R. L. *Tetrahedron Lett.* **1987**, *28*, 1123.

8. Evans, D. A.; Morrissey, M. M.; Dorow, R. L. *J. Am. Chem. Soc.* **1985**, *107*, 4346.

9. Evans, D. A.; Ennis, M. D.; Le, T.; Mandel, N.; Mandel, G. *J. Am. Chem. Soc.* **1984**, *106*, 1154.

10. Evans, D. A.; Britton, T. C. *J. Am. Chem. Soc.* **1987**, *109*, 6881.

Appendix

Chemical Abstracts Nomenclature (Collective Index Number); (Registry Number)

(S)-4-(Phenylmethyl)-2-oxazolidinone: 2-Oxazolidinone, 4-(phenylmethyl)-, (S)- (11); (90719-32-7)

(S)-Phenylalanol: 1-Propanol, 2-amino-3-phenyl-, L- or (S)-(-)- (8); Benzenepropanol, 2-amino-, (S)- (9); (3182-95-4)

(S)-Phenylalanine: Alanine, phenyl-, L- (8); L-Phenylalanine (9); (63-91-2)

DIASTEREOSELECTIVE ALDOL CONDENSATION

USING A CHIRAL OXAZOLIDINONE AUXILIARY:

(2S*,3S*)-3-HYDROXY-3-PHENYL-2-METHYLPROPANOIC ACID

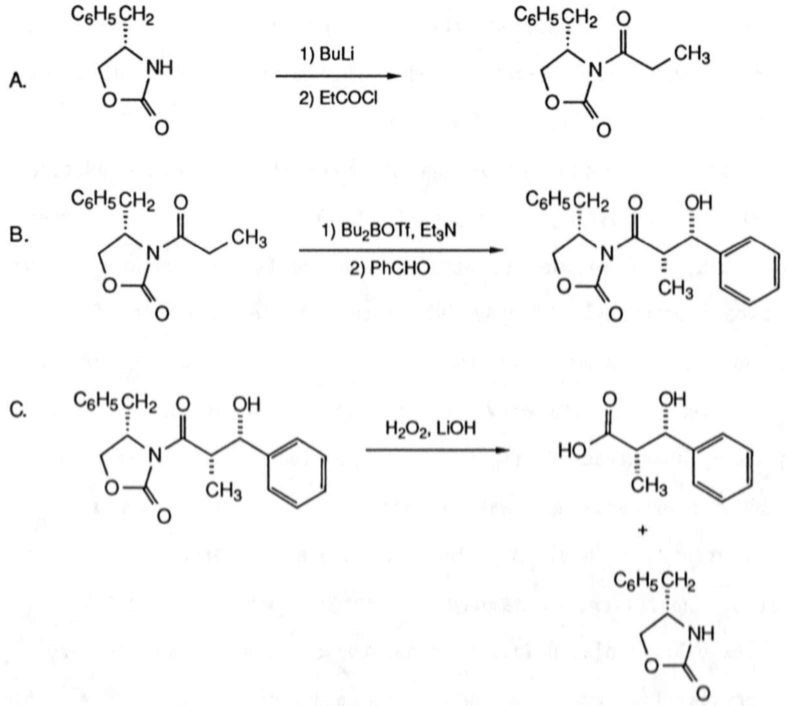

A.

B.

C.

Submitted by James R. Gage and David A. Evans.[1]

Checked by Donald T. DeRussy and Leo A. Paquette.

1. Procedure

A. (S)-3-(1-Oxopropyl)-4-(phenylmethyl)-2-oxazolidinone. A dry, 500-mL flask equipped with a magnetic stirring bar is charged with 17.7 g (0.100 mol) of (S)-4-(phenylmethyl)-2-oxazolidinone,[2] capped with a rubber septum, and flushed with nitrogen. Anhydrous tetrahydrofuran, 300 mL (Note 1), is then added to the flask via cannula, and the resulting solution is cooled to -78°C in an acetone-dry ice bath. A solution of 68.3 mL (0.101 mol) of 1.47 M butyllithium in hexane (Note 2) is transferred via cannula first to a dry, septum-stoppered, 100-mL graduated cylinder with a ground glass joint, then to the reaction flask over a 10-min period. The solution may turn yellow and slightly cloudy. Freshly distilled propionyl chloride (9.6 mL, 0.11 mol, Note 3) is added in one portion by syringe after completion of the addition of butyllithium. The resulting clear, nearly colorless solution is stirred for 30 min at -78°C, then allowed to warm to ambient temperature over a 30-min period. Excess propionyl chloride is quenched by the addition of 60 mL of saturated aqueous ammonium chloride. The bulk of the tetrahydrofuran and hexane is removed on a rotary evaporator (bath temp. ca. 25-30°C), and the resulting slurry is extracted with two 80-mL portions of dichloromethane. The combined organic extracts are washed with 75 mL of an aqueous 1 M sodium hydroxide solution and 75 mL of brine, dried over anhydrous sodium sulfate, and filtered. The solvent is removed by rotary evaporation, and the residue, a light yellow oil, is placed in a refrigerator overnight to crystallize. The resulting crystalline solid is pulverized and triturated with a minimum quantity of cold hexane. After filtration and drying 21.2-22.3 g (91-96%) of the desired product is obtained as a colorless crystalline solid, mp 44-46°C (Notes 4 and 5).

B. *The boron aldol reaction*. Into a dry, 2-L flask equipped with a large magnetic stirring bar is introduced 21.2 g (0.091 mol) of the acylated oxazolidinone. The flask is sealed with a rubber septum and swept with nitrogen. The solid is dissolved in 200 mL of anhydrous dichloromethane (Note 6), which is introduced via syringe. A thermometer is inserted through the

84

rubber septum, and the contents of the flask are cooled to 0°C with an ice bath. To this cooled solution is added via syringe 27 mL (0.107 mol) of dibutylboron triflate followed by 16.7 mL (0.120 mol) of triethylamine (Note 7) dropwise at such a rate as to keep the internal temperature below +3°C. The solution may turn slightly yellow or green during the dibutylboron triflate addition, and then to light yellow when triethylamine is added. The ice bath is then replaced with a dry ice-acetone bath (Note 8). When the internal temperature drops below -65°C, 10.3 mL (0.101 mol) of freshly distilled benzaldehyde is added over a 5-min period via syringe. The solution is stirred for 20 min in the dry ice-acetone bath, then for 1 hr at ice bath temperature. The reaction mixture is quenched by the addition of 100 mL of a pH 7 aqueous phosphate buffer and 300 mL of methanol. To this cloudy solution is added by syringe 300 mL of 2:1 methanol-30% aqueous hydrogen peroxide at such a rate as to keep the internal temperature below +10°C. After the solution is stirred for 1 additional hr, the volatile material is removed on a rotary evaporator at a bath temperature of 25-30°C. The resulting slurry is extracted with three 500-mL portions of diethyl ether. The combined organic extracts are washed with 500 mL of 5% aqueous sodium bicarbonate and 500 mL of brine, dried over anhydrous magnesium sulfate, filtered, and concentrated on a rotary evaporator, to afford 35-36 g of a white solid (Note 9). The unpurified aldol adduct has a diastereomeric purity of >97% as determined by gas chromatography (Note 10). The solid is recrystallized from ca. 500 mL of 1:2 ethyl acetate-hexane, to yield 25.8 g (84%) of the desired aldol adduct, mp 92-93°C (Note 12). The mother liquor is purified by flash chromatography (column dimensions: 8 cm x 30 cm) with flash-grade silica gel (Note 13).[3] Upon elution with 25% ethyl acetate-hexane, an additional 2.8 g (9%) of diastereomerically pure material is obtained.

C. Chiral auxiliary removal. A 500-mL flask fitted with a magnetic stirring bar is charged with 8.48 g (0.025 mol) of the aldol adduct and 125 mL of 4:1 tetrahydrofuran-distilled water. The flask is sealed with a rubber septum, purged with nitrogen, and cooled to 0°C in an ice bath. To this solution is added via syringe 10.2 mL (0.100 mol) of 30% aqueous hydrogen peroxide (Note 14) over a 5-min period, followed by 0.96 g (0.040 mol) of lithium hydroxide in 50 mL of distilled water. Some gas evolves from the clear solution. After the solution is stirred for 1 hr, the septum is removed, and 12.6 g (0.100 mol) of sodium sulfite in 75 mL of distilled water is added. The bulk of the tetrahydrofuran is removed on a rotary evaporator at a bath temperature of 25-30°C, and the resulting mixture (pH 12-13) is extracted with three 100-mL portions of dichloromethane to remove the oxazolidinone auxiliary. The aqueous layer is cooled in an ice bath and acidified to pH 1 by the addition of an aqueous 6 M hydrochloric acid solution. The resulting cloudy solution containing the β-hydroxy acid is then extracted with five 100-mL portions of ethyl acetate. The combined ethyl acetate extracts are dried over anhydrous magnesium sulfate, filtered, and concentrated, affording 5.1 g of a white crystalline solid, which is dissolved in approximately 200 mL of an aqueous 5% sodium bicarbonate solution. This solution is extracted with two 100-mL portions of dichloromethane and then acidified and extracted with ethyl acetate as before. The combined dichloromethane extracts are dried over anhydrous magnesium sulfate, filtered, and concentrated by rotary evaporation to afford 4.35 g (99%) of the oxazolidinone auxiliary as a white crystalline solid. This solid is recrystallized from 50 mL of 2:1 ethyl acetate-hexane to give 3.95 g (89%) of the recovered oxazolidinone as white crystals, mp 85-87°C. The combined ethyl acetate extracts are dried over anhydrous magnesium sulfate, filtered, and concentrated to afford 4.50 g (100%) of the desired hydroxy acid as a white crystalline solid, which is recrystallized from ca. 20 mL of carbon tetrachloride to give 4.00-4.03 g (89-90%) of pure (2S*,3S*)-3-hydroxy-3-phenyl-2-methylpropanoic acid, mp 89.5-90°C (Note 15).

2. Notes

1. Reagent grade tetrahydrofuran was purchased from Fisher Scientific Company and either freshly distilled from sodium metal and benzophenone or dried at least 3 days over activated Linde 4 Å molecular sieves before use in reaction A. It was used as received for reaction C.

2. Butyllithium in hexane was purchased from Aldrich Chemical Company, Inc. and titrated prior to use.[4]

3. Propionyl chloride (d, 1.065) was obtained from Aldrich Chemical Company, Inc., and distilled prior to use.

4. Trituration by the checkers gave 21.2-22.3 g (91-96%) of acylated product of somewhat higher purity: mp 45-46°C; $[\alpha]_D^{22}$ +99.5° (ethanol, c 1.01). Alternatively, the acylated oxazolidinone can be isolated by distillation (Kugelrohr, 125°C, 12 mm). Isolated yields are 97-99%.

5. The product has the following spectroscopic properties: IR (solution in dichloromethane) cm^{-1}: 3030, 2980, 1780, 1705, 1455, 1385, 1245, 1210, 1080; ^1H NMR (CDCl$_3$) δ: 1.2 (t, 3 H, J = 7.2, CH$_3$), 2.8 (dd, 1 H, J = 13.3, 9.6, CH$_2$C$_6$H$_5$), 2.9 (m, 2 H, CH$_2$CH$_3$), 3.3 (dd, 1 H, J = 13.4, 3.3, CH$_2$C$_6$H$_5$), 4.1 (m, 2 H, CHCH$_2$O), 4.7 (m, 1 H, NCH), 7.1-7.5 (m, 5 H, ArH); $[\alpha]_D$ +92.9° (ethanol, c 1.01).

6. Dichloromethane was distilled from calcium hydride.

7. Dibutylboron triflate was prepared according to the method of Mukaiyama.[5] It is also available from Aldrich Chemical Company, Inc. as a solution in dichloromethane or diethyl ether, but results with this material are inconsistent. It should be used within 2 weeks of preparation or after redistillation. Triethylamine (Fisher Scientific Company) was distilled from calcium hydride immediately prior to use.

8. The entire reaction can be carried out at 0°C if desired. The ratio of diastereomers in the unpurified product mixture falls slightly to 97.6:0.2:2.2 (Note 10).

9. The checkers isolated a colorless viscous oil which crystallized upon addition of 1:2 ethyl acetate-hexane. Care must be taken to avoid an excess of hexane, since oiling of the product can occur under these circumstances.

10. Diastereomer ratios were determined by gas chromatography. Since the aldol adduct undergoes retroaldol reaction on the column, it must be silylated prior to injection. Approximately 5 mg of the crude adduct is filtered through a short plug of silica gel to remove any trace metals. The material is taken up into 1-2 mL of dichloromethane in a 2-mL flask or small test tube. To this solution are added 4-5 drops of N,N-diethyl-1,1,1-trimethylsilylamine and a small crystal of 4-(N,N-dimethylamino)pyridine (Note 11). The solution is stirred for 2 hr and injected directly onto the column. (Column conditions: 30 m x 0.32 mm fused silica column coated with DB 5, 14 psi hydrogen carrier gas, oven temperature 235°C).

11. N,N-Diethyl-1,1,1-trimethylsilylamine and 4-(N,N-dimethylamino)-pyridine were purchased from Aldrich Chemical Company, Inc.

12.   The product has the following spectroscopic characteristics: IR (solution in dichloromethane) cm^{-1}:   3520, 3040, 2980, 1780, 1695, 1455, 1385, 1210, 1110; ^{1}H NMR (CDCl$_3$) δ: 1.2 (d, 3 H, J = 7.0, CH$_3$), 2.8 (dd, 1 H, J = 13.4, 9.5, 1 H CH$_2$C$_6$H$_5$), 3.1 (d, 1 H, J = 2.7, OH), 3.3 (dd, 1 H, J = 13.4, 3.4, CH$_2$C$_6$H$_5$), 4.1 (m, 3 H, CHCH$_2$O, CHCH$_3$), 4.6 (m, 1 H, NCH), 5.1 (m, 1 H, HOCH), 7.1-7.5 (m, 10 H, ArH); [α]$_D$ +75.7° (dichloromethane, c 1.00).

13.   Kieselgel 60 was purchased from EM Science, Cherry Hill, NJ, an affiliate of E. Merck, Darmstadt.

14.   Hydrogen peroxide was obtained from Mallinckrodt, Inc.

15.   The following spectroscopic characteristics were observed: IR (solution in dichloromethane) cm^{-1}:   3600, 3400-2300 broad hump, 3040, 3000, 1710, 1455, 1230; ^{1}H NMR (CDCl$_3$) δ: 1.2 (d, 3 H, J = 7.1, CH$_3$), 2.9 (m, 1 H, CHCH$_3$), 5.2 (d, 1 H, J = 3.9, C$_6$H$_5$CH), 7.2-7.4 (m, 5 H, ArH); [α]$_D^{22}$ -26.4° (CH$_2$Cl$_2$, c 1.04).   No epimerization was detected by NMR.

3. Discussion

This procedure demonstrates a particularly effective method for controlling the relative and absolute stereochemistry of the aldol reaction. It is quite general in scope.[6]   Alkyl-, aryl, and α,β-unsaturated aldehydes all give good results. In addition to chiral propionates,[7] a range of related aldol reactions may be carried out.   For example, the analogous aldol reactions of thioalkyl,[7] benzyloxy,[8] or haloacetate,[9] as well as succinate-[7] and crotonate-derived[10] carboximides, have been reported.

In addition to the high levels of asymmetric induction, two other attractive features of this sequence of reactions warrant comment. First, both acylation and hydrolysis of the chiral auxiliary are facile, high yield reactions. Second, we have recently found that the lithium hydroperoxide hydrolysis protocol described in Part C is the method of choice for the *deacylation* process. This reagent exhibits remarkable regioselectivity for attack at the desired exocyclic acyl carbonyl moiety.[11]

1. Department of Chemistry, Harvard University, Cambridge, MA 02138.

2. Gage, J. R.; Evans, D. A. *Org. Synth.* **1989**, 00, 000.

3. Still, W. C.; Kahn, M.; Mitra, A. *J. Org. Chem.* **1978**, *43*, 2923.

4. Jones, R. G.; Gilman, H. *Org. React.* **1951**, *6*, 339-366.

5. Inoue, T.; Mukaiyama, T. *Bull. Chem. Soc. Jpn.* **1980**, *53*, 174. This reference provides a very abbreviated experimental procedure. For a more detailed description for the preparation of dibutylboron triflate see: Evans, D. A.; Nelson, J. V.; Vogel, E.; Taber, T. R. *J. Am. Chem. Soc.* **1981**, *103*, 3099.

6. For a general review see: Evans, D. A. *Aldrichimica Acta* **1982**, *15*, 23.

7. Evans, D. A.; Bartroli, J.; Shih, T. L. *J. Am. Chem. Soc.* **1981**, *103*, 2127.

8. Evans, D. A.; Bender, S. L. *Tetrahedron Lett.* **1986**, *27*, 799.

9. Evans, D. A.; Sjogren, E. B.; Weber, A. E.; Conn, R. E. *Tetrahedron Lett.* **1987**, *28*, 39.

10. Evans, D. A.; Sjogren, E. B.; Bartroli, J.; Dow, R. L. *Tetrahedron Lett.* **1986**, *27*, 4957.

11. Evans, D. A.; Britton, T. C.; Ellman, J. A. *Tetrahedron Lett.* **1987**, *28*, 6141.

Appendix

Chemical Abstracts Nomenclature (Collective Index Number); (Registry Number)

3-(1-Oxopropyl)-4-(S)-phenylmethyl-2-oxazolidinone: 2-Oxazolidinone,
3-(1-oxopropyl)-4-(phenylmethyl)-, (S)- (11); (101711-78-8)

(S)-4-(Phenylmethyl)-2-oxazolidinone: 2-Oxazolidinone, 4-(phenylmethyl)-,
(S)- (11); (90719-32-7)

Propionyl chloride (8); Propanoyl chloride (9); (79-03-8)

Dibutylboron triflate: Methanesulfonic acid, trifluoro-, anhydride with
dibutylborinic acid (9); (60669-69-4)

Benzaldehyde (8,9); (100-52-7)

N,N-Diethyl-1,1,1-trimethylsilylamine: Silylamine, N,N-diethyl-1,1,1-
trimethyl- (8); Silanamine, N,N-diethyl-1,1,1-trimethyl- (9); (996-50-9)

4-(N,N-Dimethylamino)pyridine: Pyridine, 4-(dimethylamino)- (8);
4-Pyridinamine, N,N-dimethyl- (9); (1122-58-3)

1,4-DI-O-ALKYL THREITOLS FROM TARTARIC ACID:

1,4-DI-O-BENZYL-L-THREITOL

(2,3-Butanediol, 1,4-bis(phenylmethoxy)- [S-(R*,R*)]-)

A.

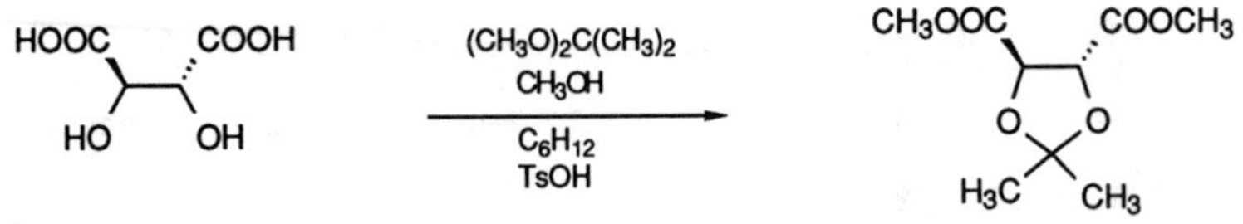

$$HOOC \quad COOH$$
$$HO \quad OH$$

$$\xrightarrow[\substack{C_6H_{12} \\ TsOH}]{\substack{(CH_3O)_2C(CH_3)_2 \\ CH_3OH}}$$

$$CH_3OOC \quad COOCH_3$$
$$O \quad O$$
$$H_3C \quad CH_3$$

B.

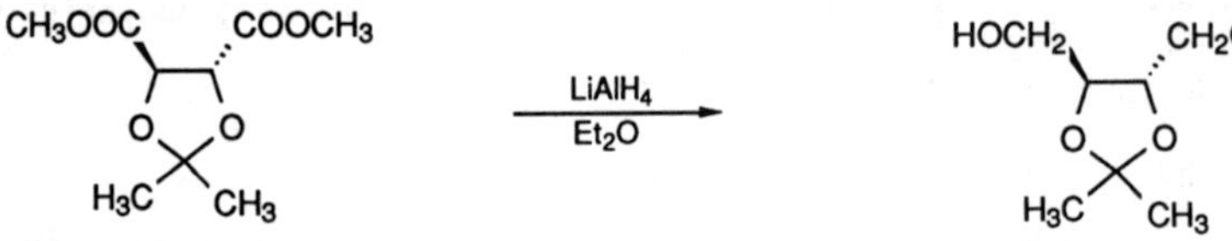

$$CH_3OOC \quad COOCH_3$$
$$O \quad O$$
$$H_3C \quad CH_3$$

$$\xrightarrow[\substack{Et_2O}]{LiAlH_4}$$

$$HOCH_2 \quad CH_2OH$$
$$O \quad O$$
$$H_3C \quad CH_3$$

C.

$$HOCH_2 \quad CH_2OH$$
$$O \quad O$$
$$H_3C \quad CH_3$$

$$\xrightarrow[\substack{THF}]{\substack{NaH \\ PhCH_2Br}}$$

$$PhCH_2OCH_2 \quad CH_2OCH_2Ph$$
$$O \quad O$$
$$H_3C \quad CH_3$$

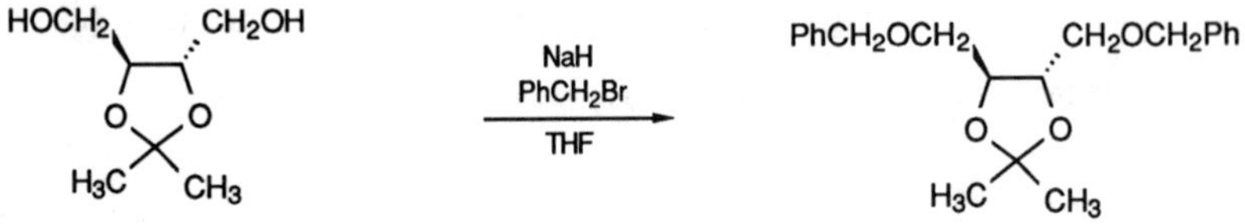

D.

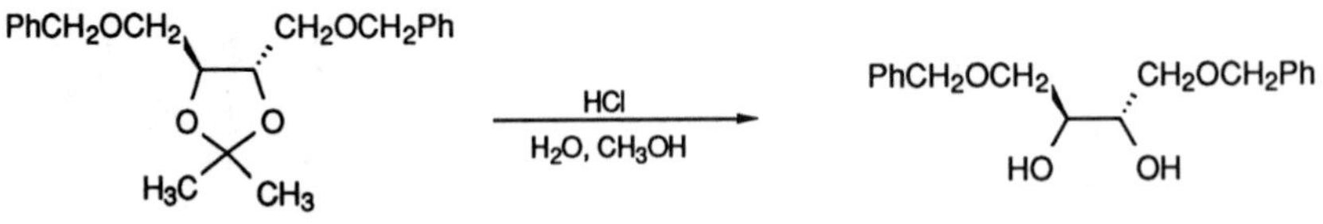

$$PhCH_2OCH_2 \quad CH_2OCH_2Ph$$
$$O \quad O$$
$$H_3C \quad CH_3$$

$$\xrightarrow[\substack{H_2O, CH_3OH}]{HCl}$$

$$PhCH_2OCH_2 \quad CH_2OCH_2Ph$$
$$HO \quad OH$$

Submitted by Eugene A. Mash, Keith A. Nelson, Shawne Van Deusen, and Susan B. Hemperly.[1]

Checked by Peter D. Theisen and Clayton H. Heathcock.

1. Procedure

A. Dimethyl 2,3-O-isopropylidene-L-tartrate (Note 1). In a 1-L, one-necked, round-bottomed flask fitted with a reflux condenser and a large magnetic stirring bar under argon, a mixture of L-tartaric acid (101 g, 0.673 mol) (Note 2), 2,2-dimethoxypropane (190 mL, 161 g, 1.54 mol) (Note 3), methanol (40 mL) (Note 4), and p-toluenesulfonic acid monohydrate (0.4 g, 2.1 mmol) (Note 5) is warmed on a steam bath with occasional swirling until a dark red homogeneous solution is obtained (Note 6). Additional 2,2-dimethoxy-propane (95 mL, 80.5 g, 0.77 mol) and cyclohexane (450 mL) (Note 7) are added and the flask is fitted with a 30-cm Vigreux column and a variable reflux distilling head. The mixture is heated to reflux with internal stirring and the acetone-cyclohexane and methanol-cyclohexane azeotropes are slowly removed (Note 8). Additional 2,2-dimethoxypropane (6 mL, 5.1 g, 49 mmol) is then added and the mixture heated under reflux for 15 min (Note 9). After the mixture has cooled to room temperature, anhydrous potassium carbonate (1 g, 7.2 mmol) is added and the mixture is stirred until the reddish color has abated (Note 10). Volatile material is removed under reduced pressure (water aspirator) and the residue is fractionally distilled under vacuum to afford the product as a pale yellow oil, bp 94-101°C (0.5 mm); 125-135 g (0.57-0.62 mmol, 85-92% yield) (Notes 11 and 12).

B. 2,3-Di-O-isopropylidene-L-threitol (Note 13). In a dry, 2-L, three-necked, round-bottomed flask equipped with a 500-mL pressure-equalized addition funnel, reflux condenser, thermometer, and a large magnetic stirring bar, is suspended lithium aluminum hydride (36 g, 0.95 mol) (Note 14) in diethyl ether (600 mL) (Note 15) under argon. The mixture is stirred and heated to reflux for 30 min. Heating is discontinued while a solution of

93

dimethyl 2,3-0-isopropylidene-L-tartrate (123 g, 0.564 mol) in diethyl ether (300 mL) (Note 15) is added dropwise over 2 hr. Heating is then resumed and the mixture is refluxed for an additional 3 hr. The mixture is cooled to 0-5°C (Note 16) and *cautiously* treated with water (36 mL), 4 N sodium hydroxide solution (36 mL), and water (112 mL) (Note 17). The mixture is then stirred at room temperature until the grey color of unquenched lithium aluminum hydride has completely disappeared (Note 18). The mixture is filtered on a Büchner funnel and the inorganic precipitate is extracted with ether in a Soxhlet apparatus (Note 19). The combined ethereal extracts are dried over anhydrous magnesium sulfate and filtered, and volatile material is removed under reduced pressure (water aspirator). The residue is fractionally distilled under vacuum to afford the product as a colorless to pale yellow oil, bp 94-106°C (0.4 mm); 50.2-60.3 g (0.31-0.37 mmol, 54-66% yield) (Notes 20 and 21).

C. 1,4-Di-O-benzyl-2,3-di-O-isopropylidene-L-threitol. In a 2-L, three-necked, round-bottomed flask, equipped with a large magnetic stirring bar under argon, is placed fresh sodium hydride (33.6 g of a 55% dispersion in oil; 18.5 g, 0.77 mol) (Note 22). The oil is removed by washing with hexanes (3 x 100 mL) (Notes 23 and 24). The flask is fitted with a 500-mL pressure-equalized addition funnel, a reflux condenser, and a stopper. Tetrahydrofuran (250 mL) (Note 25) is added under argon. A solution of 2,3-di-O-isopropylidene-L-threitol (55 g, 0.34 mol) in tetrahydrofuran (250 mL) is then added dropwise with stirring at room temperature (Note 26). Benzyl bromide (91 mL, 130.8 g 0.76 mol) (Note 27) is then added dropwise via the addition funnel (Note 28). After stirring for 12 hr at room temperature, the mixture is heated at reflux for 2 hr, cooled in an ice bath, and quenched by addition of water until a clear solution results. Tetrahydrofuran is removed under reduced pressure (water aspirator); the residue is diluted with water (500 mL) and extracted with diethyl ether (3 x 500 mL). The extracts are combined, dried over anhydrous magnesium sulfate, and filtered. Removal of volatile material under reduced pressure (water aspirator) gives crude 1,4-di-0-benzyl-2,3-0-isopropylidene-L-threitol as an oil (115-116 g).

D. *1,4-Di-0-benzyl-L-threitol.* The crude ketal is dissolved in methanol (300 mL), 0.5 N hydrochloric acid (30 mL) is added, and the resulting mixture is heated to reflux. Acetone and methanol are slowly distilled off (Note 29). Additional methanol (50 mL) and 0.5 N hydrochloric acid (20 mL) are added and the mixture is kept at room temperature until ketal hydrolysis is complete. The mixture is diluted with saturated sodium bicarbonate solution (500 mL) and extracted with ether (3 x 500 mL). The ether extracts are combined, dried over anhydrous magnesium sulfate, and filtered. Removal of volatile material under reduced pressure gives crude 1,4-di-0-benzyl-L-threitol as a pale yellow solid. This solid is recrystallized twice from chloroform/hexanes, to provide 59-65 g (195-215 mmol, 57-63% yield) of pure diol, mp 54-55°C (Notes 30 and 31). Concentration of the mother liquors from the recrystallizations gives a yellow solid which is chromatographed on 70-230 mesh silica gel 60 (500 g) (Note 32), and eluted with 50% ethyl acetate/hexanes, to provide an additional 20-26 g (66-86 mmol, 19-25% yield) of diol, mp 56-57°C (Note 33).

2. Notes

1. Dimethyl 2,3-0-isopropylidene-L-tartrate is also commercially available from Fluka Chemical Corporation.

2. L-Tartaric acid, 99+%, mp 170-172°C, $[\alpha]_D^{20}$ +12.4° (water, c 20), from Aldrich Chemical Company, Inc., was used as obtained.

3. 2,2-Dimethoxypropane, 98%, from Aldrich Chemical Company, Inc., was distilled before use. The checkers used this material directly from the bottle without adverse effects.

4. Methanol was distilled from sodium methoxide before use. The checkers used absolute methanol from Fisher Chemical Company directly from the bottle without adverse effects.

5. p-Toluenesulfonic acid monohydrate, 99%, from Aldrich Chemical Company, Inc., was used as obtained.

6. This normally requires 1-2 hr.

7. Reagent grade cyclohexane, from MCB Manufacturing Chemists Inc., was used as obtained.

8. Removal over a 2-day period (10-15 mL/hr) is satisfactory. After approximately 600 mL of distillate has been collected, the temperature at the solvent head is approximately 79°C.

9. The checkers omitted this final addition and 15-min reflux period without adverse effects.

10. This normally occurs within 1-2 hr, leaving a yellow solution.

11. This product was $\geq$ 88% pure based on recovery of an analytical sample from chromatography on silica gel 60 eluted with 30% ethyl acetate/hexanes. Physical properties and spectral data are as follows: $[\alpha]_D^{24}$ -42.6° (CHCl$_3$, c 5.1), lit. $[\alpha]_D^{20}$ -49.4° (neat);[2] IR (neat) cm^{-1}: 2992, 2956, 1759, 1438, 1384, 1213, 1111, 1013, 858, 749; ^{1}H NMR (CDCl$_3$) δ: 1.49 (s, 6 H), 3.83 (s, 6 H), 4.81 (s, 2 H); ^{13}C NMR (CDCl$_3$) δ: 25.98, 52.42, 76.68, 113.49, 169.75. The checkers found this material to be 91-94% pure by gas chromatography.

12. Dimethyl 2,3-0-isopropylidene-D-threitol was prepared similarly.

13. 2,3-0-Isopropylidene-L-threitol is also commercially available from Aldrich Chemical Company, Inc. and from Fluka Chemical Corporation.

14. Lithium aluminum hydride, 95+%, from Aldrich Chemical Company, Inc., was used as obtained.

15. Diethyl ether was distilled from sodium immediately prior to use.

16. An ice/salt or ice/acetone bath is employed.

17. Dropwise addition via a funnel is recommended. During the quenching procedure, stirring becomes difficult for a time and manual swirling may be necessary. Use of a stirrer with a powerful magnet is recommended.

18. *Cautious* scraping of the sides of the flask to expose isolated pockets of unquenched lithium aluminum hydride may be expeditious at this point.

19. The checkers obtained a greater yield by carrying out the Soxhlet extraction with tetrahydrofuran instead of ether.

20. This product was ≥ 80% pure based on recovery of an analytical sample from chromatography on silica gel 60 eluted with 80% ethyl acetate/hexanes. Physical properties and spectral data are as follows: $[\alpha]_D^{24}$ +2.78° (CHCl$_3$, c 4.67), lit. $[\alpha]_D^{20}$ +4.1° (CHCl$_3$, c 5);[3] IR (neat) cm^{-1}: 3413, 2987, 2934, 1455, 1372, 1218, 1166, 1057, 986, 882, 844, 801, 756; ^{1}H NMR (CDCl$_3$) δ: 1.42 (s, 6 H), 3.73 (m, 6 H), 3.94 (m, 2 H); ^{13}C NMR (CDCl$_3$) δ: 26.75, 62.06, 78.32, 109.08. The checkers found this material to be 95-97% pure by gas chromatography.

21. 2,3-Di-0-isopropylidene-D-threitol was prepared similarly.

22. Sodium hydride, 55-60% disperison in mineral oil, from Alfa Products, Morton/Thiokol Inc., was used as obtained.

23. Hexanes were distilled from calcium hydride prior to use.

97

24. The hexane washes can be decanted into a large beaker containing isopropyl alcohol and dry ice. The last traces of hexanes can be removed under vacuum, followed by reintroduction of an argon atmosphere.

25. Tetrahydrofuran was distilled from sodium immediately prior to use.

26. Toward the end of this addition stirring becomes increasingly difficult. Use of a stirrer with a powerful magnet is recommended.

27. Benzyl bromide from Fluka Chemical Corporation was used as obtained.

28. If magnetic stirring is impossible at this point, manual swirling of the flask may be necessary for a time. As the alkylation proceeds, the mixture becomes less viscous.

29. This requires 3-5 hr. Progress of this hydrolysis can be monitored by thin layer chromatography on 0.25-mm silica gel 60 plates eluted with 50% ethyl acetate/hexanes; R_f ketal 0.59, R_f diol 0.21.

30. Physical properties and spectral data for 1,4-di-0-benzyl-L-threitol are as follows: $[\alpha]_D^{24}$ -5.85° (CHCl$_3$, c, 5.15), lit. $[\alpha]_D^{25}$ -5.0 (CHCl$_3$, c 5);[4] IR (CHCl$_3$) cm^{-1}: 3562, 3065, 3014, 2867, 1495, 1453, 1361, 1233, 1101, 1027; ^{1}H NMR (CDCl$_3$) δ: 2.93 (br s, 2 H), 3.54-3.60 (m, 4 H), 3.83-3.87 (m, 2 H), 4.47-4.57 (m, 4 H), 7.23-7.36 (m, 10 H); ^{13}C NMR (CDCl$_3$) δ: 70.45, 71.85, 73.45, 127.70, 128.37, 137.66.

31. 1,4-Di-0-benzyl-D-threitol was prepared similarly; $[\alpha]_D^{24}$ +6.16° (CHCl$_3$, c 3.83).

32. E. Merck 70-230 mesh silica gel 60 from Curtin-Matheson Scientific was employed.

33. The checkers found that the solid obtained by recrystallization from chloroform/hexanes occludes a large amount of solvent. To obtain pure, dry material, it is necessary to press the moist solid while it is still on the Büchner funnel and then to dry it under vacuum (room temperature, 0.05 mm, 12-

18 hr). Only 35 g of pure material was obtained in this manner. Repetition of the process with the mother liquors yielded another 35 g of material. The remaining product (ca. 20 g) was obtained by chromatography. Recrystallization from ethyl acetate/hexanes gave a product that is easier to dry.

3. Discussion

Homochiral molecules readily available from inexpensive sources are useful synthetic building blocks and chiral auxiliaries. 1,4-Di-0-benzyl-L-threitol has been used in construction of homochiral crown ethers that are useful as enzyme model systems.[5] Topologically controlled diastereoselective delivery of the Simmons-Smith reagent for 2-cycloalken-1-one 1,4-di-0-benzyl-L-threitol ketals was recently reported.[6]

A number of other enantioselective processes are known to depend on homochiral acetal or ketal participation.[7] Diols used in these reactions include tartrate esters, tartramides, propanediols, butanediols, and pentanediols. 1,4-Di-0-benzyl-L-threitol may prove superior to other diols since: (a) it can be produced inexpensively in quantity in either enantiomeric form; (b) it is an amorphous solid; (c) it contains a UV chromophore, making derivatives easy to monitor; (d) it can be introduced directly or via transketalization; (e) it provides "functionalized arms" which can be chemically manipulated after ketalization.

The preparation of 1,4-di-0-benzyl-L-threitol described here from L-tartaric acid via 2,3-0-isopropylidene-L-threitol is adapted from work by Carmack,[2] Feit,[3] and Inouye.[4] This general route has been employed by the

243242

99

submitters and by others for the production of a number of synthetically useful L-threitol derivatives (Table). The corresponding D-threitol derivatives are as easily prepared from commercially available D-tartaric acid.

1. Department of Chemistry, University of Arizona, Tucson, AZ 85721.

2. Carmack, M.; Kelley, C. J. *J. Org. Chem.* **1968**, *33*, 2171-2173.

3. Feit, P. W. *J. Med. Chem.* **1964**, *7*, 14-17.

4. Ando, N.; Yamamoto, Y.; Oda, J.; Inouye, Y. *Synthesis* **1978**, 688-690.

5. (a) Sasaki, S.; Kawasaki, M., Koga, K. *Chem. Pharm. Bull.* **1985**, *33*, 4247-4266; (b) Fyles, T. M.; McGavin, C. A.; Whitfield, D. M. *J. Org. Chem.* **1984**, *49*, 753-761; (c) Helgeson, R. C.; Weisman, G. R.; Toner, J. L.; Tarnowski, T. L.; Chao, Y.; Mayer, J. M.; Cram, D. J. *J. Am. Chem. Soc.* **1979**, *101*, 4928-4941.

6. (a) Mash, E. A.; Nelson, K. A. *J. Am. Chem. Soc.* **1985**, *107*, 8256-8258; (b) Mash, E. A.; Nelson, K. A. *Tetrahedron Lett.* **1986**, *27*, 1441-1444; (c) Nelson, K. A.; Mash, E. A. *J. Org. Chem.* **1986**, *51*, 2721-2724.

7. (a) Johnson, W. S.; Harbert, C. A.; Stipanovic, R. D. *J. Am. Chem. Soc.* **1968**, *90*, 5279; (b) Bartlett, P. A.; Johnson, W. S.; Elliott, J. D. *J. Am. Chem. Soc.* **1983**, *105*, 2088; (c) Choi, V. M. F.; Elliott, J. D.; Johnson, W. S. *Tetrahedron Lett.* **1984**, *25*, 591; (d) Johnson W. S.; Chan, M. F. *J. Org. Chem.* **1985**, *50*, 2598; (e) Mori, A.; Fujiwara, J.; Maruoka, K.; Yamamoto, H. *Tetrahedron Lett.* **1983**, *24*, 4581; (f) Mori, A.; Fujiwara, J.; Maruoka, K.; Yamamoto, H. *J. Organomet. Chem.* **1985**, *285*, 83; (g) Mori, A.; Ishihara, K.; Yamamoto, H. *Tetrahedron Lett.* **1986**, *27*, 987; (h) Fujiwara, J.; Fukutani, Y.; Hasegawa, M.; Maruoka, K.; Yamamoto, H. *J. Am. Chem. Soc.* **1984**, *106*, 5004; (i) Alexakis, A.; Mangeney, P.;

Normant, J. F. *Tetrahedron Lett.* **1985**, *26*, 4197; (j) Ghribi, A.;
Alexakis, A.; Normant, J. F. *Tetrahedron Lett.* **1984**, *25*, 3083; (k)
Mashraqui, S. H.; Kellogg, R. M. *J. Org. Chem.* **1984**, *49*, 2513; (l)
McNamara, J. M.; Kishi, Y. *J. Am. Chem. Soc.* **1982**, *104*, 7371; (m)
Richter, W. J. *J. Org. Chem.* **1981**, *46*, 5119; (n) Tamura, Y.; Kondo, H.;
Annoura, H; Takeuchi, R.; Fujioka, H. *Tetrahedron Lett.* **1986**, *27*, 81; (o)
Imwinkelried, R.; Seebach, D. *Angew. Chem., Intern. Ed. Engl.* **1985**, *24*,
765; (p) Winstead, R. C. Simpson, T. H.; Lock, G. A.; Schiavelli, M. D.;
Thompson, D. W. *J. Org. Chem.* **1986**, *51*, 275.

8. Seebach, D.; Kalinowski, H.-O.; Bastani, B.; Crass, G.; Daum, H.; Dörr,
   H.; DuPreez, N. P.; Ehrig, V.; Langer, W.; Nüssler, C.; Oei, H.-A.;
   Schmidt, M. *Helv. Chim. Acta* **1977**, *60*, 301-325.

9. Van Deusen, Shawne, unpublished results.

10. Rubin, L. J.; Lardy, H. A., Fischer, H. O. L. *J. Am. Chem. Soc.* **1952**, *74*,
    425-428.

11. Plattner, J. J.; Rapoport, H. *J. Am. Chem. Soc.* **1971**, *93*, 1758-1761.

Electrophile	Threitol Derivative	Yield, %[a]	Reference
Methyl iodide	HO OCH3 / HO OCH3	62	8
Butyl iodide	HO OBu / HO OBu	37	8
4-(Chloromethyl)biphenyl	HO OCH2C6H4Ph-p / HO OCH2C6H4Ph-p	53	9
2-(Bromomethyl)naphthalene	HO OCH2(β-naphthyl) / HO OCH2(β-naphthyl)	83	9
p-Toluenesulfonyl chloride	H3C O OTs / H3C O OTs	82-90	1, 8, 10
Methanesulfonyl chloride	H3C O OMs / H3C O OMs	75-82	2, 11

Appendix

Chemical Abstracts Nomenclature (Collective Index Number);

(Registry Number)

1,4-Di-0-benzyl-L-threitol: 2,3-Butanediol, 1,4-bis(benzyloxy)-, (2S,3S)-
(8); 2,3-Butanediol, 1,4-bis(phenylmethoxy)-, [S-(R*,R*)]- (9); (17401-06-8)

Dimethyl 2,3-0-isopropylidene-L-tartrate: 1,3-Dioxolane-4,5-dicarboxylic
acid, 2,2-dimethyl-, dimethyl ester, (4R-trans)- (9); (37031-29-1)

L-Tartaric acid: Tartaric acid, L- (8); Butanedioic acid, 2,3-dihydroxy-,
[R-(R*,R*)]- (9), (87-69-4)

2,2-Dimethoxypropane: Propane, 2,2-dimethoxy- (9); (77-76-9)

2,3-Di-0-isopropylidene-L-threitol: 1,3-Dioxolane-4,5-dimethanol,
2,2-dimethyl-, (4S-trans)- (9); (50622-09-8)

ALLYLTRIBUTYLTIN

(Stannane, tributyl-2-propenyl-)

$$CH_2{=}CHCH_2Br \quad + \quad Mg \quad \xrightarrow{\text{Et}_2O} \quad CH_2{=}CHCH_2MgBr$$

$$2\ CH_2{=}CHCH_2MgBr \quad + \quad [(n\text{-}C_4H_9)_3Sn]_2O \quad \longrightarrow \quad 2\ (n\text{-}C_4H_9)_3SnCH_2CH{=}CH_2$$

Submitted by Noreen G. Halligan and Larry C. Blaszczak.[1]

Checked by Christophe M. G. Philippo and Leo A. Paquette.

1. Procedure

Magnesium turnings (72.6 g, 3 g-atom) and 1000 mL of anhydrous ethyl ether are placed under argon in a dry, 3-L, three-necked flask equipped with a mechanical stirrer, 500-mL pressure-equalizing dropping funnel, Claisen adapter, thermometer, ice water-cooled condenser and argon inlet. The dropping funnel is charged with allyl bromide (158 mL, 1.8 mol) (Note 1) in 150 mL of anhydrous ether. Stirring is initiated and a 10-12 mL portion of the allyl bromide solution is run into the reaction flask. The resulting mixture is treated with a few crystals of iodine, whereupon a rise in temperature and clouding of the reaction mixture occurs indicating that the reaction has begun (Note 2). The remainder of the allyl bromide solution is added dropwise with continued stirring at such a rate as to maintain a gentle reflux. The addition requires approximately 1.5 hr. The mixture is then refluxed for an additional 1.5 hr.

While the Grignard solution is being refluxed, the dropping funnel is charged with bis(tributyltin) oxide (371 g, 0.62 mol) (Note 1) in 150 mL of anhydrous ether. After the reflux, heating is stopped and the bis(tri-butyltin) oxide solution is added at such a rate as to maintain a reaction temperature of 36-38°C. The addition requires approximately 1 hr. After the addition is complete, the reaction mixture is refluxed for 1.5 hr and then stirred overnight at room temperature.

The reaction mixture is cooled in a water-ice bath, and a saturated aqueous ammonium chloride solution is added at such a rate as to maintain the temperature below 35°C. Ammonium chloride solution is added in portions until addition produces no further exothermic reaction (Note 3). The supernatant solution is decanted through glass wool onto 400 g of ice in a 4-L separatory funnel. The residual solids are washed with three portions of hexane, approximately 1000 mL total, and the washes are decanted into the separatory funnel. After the phases are separated, the aqueous phase is washed with an additional 500-mL portion of hexane. The combined organic extracts are washed with 500 mL of saturated ammonium chloride, and then with 500 mL of brine. The organic layer is dried over anhydrous magnesium sulfate and filtered. Most of the solvent is removed by a rotary evaporator and the residual oil is distilled at reduced pressure using an ice water-cooled fraction cutting head. After a small forerun, approximately 390-392 g (94% of theory) is collected as a colorless oil, bp 116°C/1.6 mm (lit. 155°C/17 mm).[2]

2. Notes

1. Allyl bromide and bis(tributyltin) oxide were obtained from Aldrich Chemical Company, Inc.

2. Initiation of the reaction required about 1 min of sonication in an ordinary laboratory cleaner.

3. Approximately 190 mL of saturated aqueous ammonium chloride is required.

3. Discussion

Within the last decade, organotin chemistry has become a major source of new and highly selective reagents for effecting carbon-carbon bond formation. Transmetallation, nucleophilic substitution, stereoselective carbonyl addition, and transition metal- or radical-mediated substitution reactions have all been accomplished using allyltributyltin. Because of the broad range of selective reactivities under which the synthetically versatile allyl group may be transferred to a highly functionalized substrate, allyltin compounds have secured a position on the modern chemist's list of indispensable reagents.

Transmetallation of allyltributyltin with organolithium species[3] has been used for the generation of allyllithium solutions free of the coupling by-products which often result from reduction of allylic halides with lithium metal. These solutions may then be used directly for the preparation of Gilman reagents and other reactive modifications of the parent allyllithium.

The use of allyltributyltin in combination with a Lewis acid has been used to effect both nucleophilic substitution[4] and stereoselective carbonyl addition[5] reactions. These reactions occur with a high degree of selectivity because of the reagent's nucleophilic, completely non-basic character in the presence of a sufficiently reactive carbon electrophile. Allyltin reagents appear to be more useful than the corresponding allylsilanes for these purposes.

By far the most generally useful synthetic application of allyltributyltin is in the complementary set of transition metal- and radical-mediated substitution reactions. When the halide substrates are benzylic, allylic, aromatic or acyl, transition metal catalysis[6] is usually the method of choice for allyl transfer from tin to carbon. When the halide (or halide equivalent) substrate is aliphatic or alicyclic, radical chain conditions[7] are appropriate, as β-hydrogen elimination is generally not a problem in these cases.

Allyltriorganotin compounds have been prepared by the reaction of allyl Grignard[8] or allyllithium reagents with triorganotin halides as well as by the procedure described. This procedure is an adaptation of that used by Rosenberg[9] for the preparation of vinyltrialkyltin compounds. Allyltriorganotin compounds in which the allyl group bears complex substituents can be prepared by desulfurization of allylic sulfides, sulfoxides, or sulfones with triorganotin hydrides.[10]

1. The Lilly Research Laboratories, Lilly Corporate Center, Indianapolis, IN 46285.

2. In "Dictionary of Organometallic Compounds"; Buckingham, J., Ed.; Chapman and Hall Ltd.: New York, 1984; Vol. 2, p. 2196.

3. Seyferth, D.; Weiner, M. A. *J. Org. Chem.* **1959**, *24*, 1395.

4. Trost, B. M.; Sato, T. *J. Am. Chem. Soc.* **1985**, *107*, 719

5. Keck, G. E.; Boden, E. P. *Tetrahedron Lett.* **1984**, *25*, 265, 1879.

6. Kosugi, M.; Sasazawa, K.; Shimizu, Y.; Migita, T. *Chem. Lett.* **1977**, 301; Milstein, D.; Stille, J. K. *J. Am. Chem. Soc.* **1979**, *101*, 4992.

7. Kosugi, M.; Kurino, K.; Takayama, K.; Migita, T. *J. Organomet. Chem.* **1973**, *56*, C11; Grignon, J.; Pereyre, M. *J. Organomet. Chem.* **1973**, *61*, C33; Keck, G. E.; Yates, J. B. *J. Am. Chem. Soc.* **1982**, *104*, 5829.

8. Seyferth, D.; Weiner, M. A. *J. Org. Chem.* **1961**, *26*, 4797.

9. Rosenberg, S. D.; Gibbons, Jr., A. J.; Ramsden, H. E. *J. Am. Chem. Soc.* **1957**, *79*, 2137.

10. Ueno, Y.; Aoki, S.; Okawara, M. *J. Am. Chem. Soc.* **1979**, *101*, 5414.

Appendix
Chemical Abstracts Nomenclature (Collective Index Number)
(Registry Number)

Allyltributyltin: Stannane, allyltributyl- (8); Stannane, tributyl-2-propenyl- (9); (24850-33-7)

Allyl bromide: 1-Propene, 3-bromo- (8,9); (106-95-6)

Bis(tributyltin) oxide: Distannoxane, hexabutyl- (8,9); (56-35-9)

ALLYLIC ACETOXYLATION OF CYCLOALKENES:

2-CYCLOHEPTEN-1-YL ACETATE

(2-Cyclohepten-1-ol, acetate)

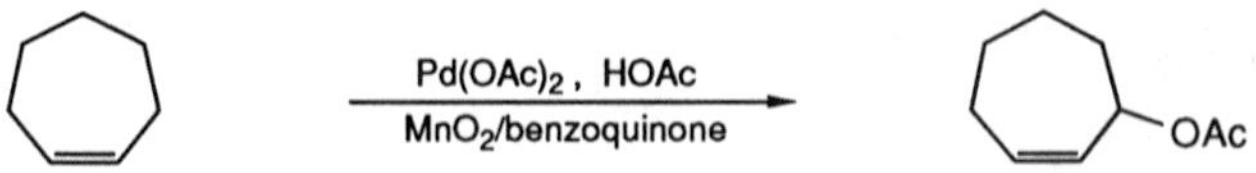

Submitted by A. Heumann,[1] B. Åkermark,[2] S. Hansson,[2] and T. Rein.[2]
Checked by Joe Guiles and Albert I. Meyers.

1. Procedure

Palladium acetate (1.12 g, 0.005 mol), benzoquinone (2.16 g, 0.02 mol), manganese dioxide (10.44 g, 0.12 mol) and anhydrous acetic acid (250 mL) (Note 1), are placed in a 1-L, round-bottomed flask equipped with a reflux condenser and magnetic stirring bar. This heterogeneous mixture is equilibrated by efficient stirring for 30-60 min. Cycloheptene (9.61 g, 0.1 mol) (Note 2) is added, and the stirring is continued at 60°C for 28 hr (Note 3). After the solution is cooled to room temperature, 250 mL of pentane/ether (1:1) is added and the mixture is stirred for another 30 min. The two-phase mixture is filtered with suction through a Buchner funnel, which contains a layer of Celite (5-10 mm). The Celite layer is washed successively with 250 mL of pentane/ether (1:1), 250 mL of water, 100 mL of pentane/ether (1:1), and 250 mL of water. After the organic phases are separated, the aqueous phase is extracted three times with 250 mL of pentane/ether (1:1). The combined organic phases are washed successively with 250 mL of water, 250 mL and then

109

100 mL of aqueous sodium hydroxide (2 N) (Note 4), 250 mL of water, and finally dried over anhydrous magnesium sulfate. After evaporation or distillation of the solvent, the product is purified by distillation (Note 5) to give 2-cyclohepten-1-yl acetate (11.25 g, 73%), bp 61-62°C (5 mm), lit.[3] bp 70°C (6 mm) (Note 6).

2. Notes

1. All the reagents used are analytical grade, commercially available products, which are used without further purification. Darkened benzoquinone was purified by sublimation. Activated grade manganese dioxide was used; however it was not shown that "activation" of manganese dioxide is necessary for the reaction.

2. Reaction conditions for other olefins are shown in Table I.

3. The time for optimized conversion has been determined by GLC for all olefins. It is crucial for all reactions to be stopped at optimum conversion, because slow decomposition of the allylic product occurs during the reaction. To obtain optimum yields one should follow the reaction by GLC. Optimized conversion is defined as: allylic acetate/allylic acetate plus remaining olefin.

4. *Caution should be observed during the alkaline washings because they are exothermic.*

5. The crude reaction products can easily be purified by distillation or by flash chromatography, with hexane/ether (95:5) as eluant.

6. The product exhibits the following NMR spectra: [1]H (200 MHz, CDCl$_3$) δ: 1.30-2.30 (m, 8 H), 2.05 (s, 3 H), 5.40 (m, 1 H), 5.65 (m, 1 H), 5.82 (m, 1 H); [13]C (50.3 MHz, CDCl$_3$) δ: 21.20, 26.43, 26.48, 28.33, 32.70, 74.13, 131.38, 133.56, 170.24.

3. Discussion

Allylic acetates are usually prepared by esterification from allylic alcohols. However, the corresponding alcohols are often only accessible by the fairly expensive hydride reduction of carbonyl compounds. Consequently, direct allylic functionalization of easily available olefins has been intensively investigated.[4] Most of these reactions involve peroxides[5] or a variety of metal salts.[6,7] However, serious drawbacks of these reactions, (e.g. toxicity of some metals, stoichiometric reaction conditions, or non-generality) may be responsible for their infrequent use for the construction of allylic alcohols or acetates.

Allylic acetoxylation with palladium(II) salts is well known;[8] however, no selective and catalytic conditions have been described for the transformation of an unsubstituted olefin. In the present system use is made of the ability of palladium acetate to give allylic functionalization (most probably via a palladium-π-allyl complex) and to be easily regenerated by a co-oxidant (the combination of benzoquinone-manganese dioxide). In contrast to copper(II) chloride (CuCl$_2$) as a reoxidant,[8] our catalyst combination is completely regioselective for alicyclic alkenes; with aliphatic substrates, evidently, both allylic positions become substituted. As yet, no allylic oxidation reagent is able to distinguish between the two allylic positions in linear olefins; this disadvantage is overcome when the allylic acetates are to be used as precursors for π-allyl complexes (for example in palladium-catalyzed substitution reactions).

1. Université d'Aix-Marseille, Faculté de St.-Jérome, UA 126,1PSOI F 13397 Marseille Cedex 13, France.

2. Royal Institute of Technology, Department of Organic Chemistry, S 100 44 Stockholm, Sweden.

3. Cope, A. C.; Liss, T. A.; Wood, G. W. *J. Am. Chem. Soc.* **1957**, *79*, 6287-6292; *Chem. Abstr.* **1958**, *52*, 6219f; Beilstein, *6/1*, EIV 202.

4. Muzart, J. *Bull. Soc. Chim. Fr.* **1986**, 65.

5. Rawlinson, D. J.; Sosnovsky, G. *Synthesis* **1972**, 1.

6. Rawlinson, D. J.; Sosnovsky, G. *Synthesis* **1973**, 567.

7. Friedrich, A. In "Methoden der Organischen Chemie," 4th ed.; Houben-Weyl, Ed.; Thieme: Stuttgart, 1975; Vol. 4, Part 1b: Hg(II) pp 89-90; Rotermund, G. W. In "Methoden der Organischen Chemie," 4th ed.; Houben-Weyl, Ed.; Thieme: Stuttgart, 1975; Vol. 4, Part 1b: Tl(III), p 127; Rotermund, G. W. In "Methoden der Organischen Chemie", 4th ed.; Houben-Weyl, Ed.; Thieme: Stuttgart, 1975; Vol. 4, Part 1b: Pb(IV) pp 220-222; Kropf, H. In "Methoden der Organischen Chemie," 4th ed.; Houben-Weyl, Ed.; Thieme: Stuttgart, 1979; Vol. 6/1a, Part 1: Pb(IV) pp 25-72.

8. Henry, P. M. "Palladium Catalyzed Oxidation of Hydrocarbons", Kluwer Academic Publ.: Boston, MA, 1979.

TABLE

ALLYLIC ACETOXYLATION OF OLEFINS

Olefin (mmol)	Pd(OAc)$_2$ mmol	MnO$_2$ mmol	[quinone] mmol	HOAc mL	Temp. °C	Time hr	Yield %	Product[a]	bp °C (mm)	Optimized Conversion (Note 3)
cyclopentene (100)	.5	120	20	250	50	16	66	(OAc product)	79-82 (58)	>95[b]
cyclohexene (100)	0.5	110	10	250	60	50	77	(OAc product)	68 (15)	95
cycloheptene (100)	5	120	20	250	60	28	73	(OAc product)	61-62 (5)	98
cyclooctene (100)	5	200	20	250	60	90	35[c]	(OAc product)	63-64 (3)	60
cyclodecene (20)	1	24	4	50	60	300	78	(OAc product)	flash (Note 5)	93
cycloundecene (20)	1	24	4	50	60	43	72[d]	(OAc product)	flash (Note 5)	77
(linear diene) (2)	0.1	2.4	0.4	5	60	72	>80[e]	(OAc products)[f]	flash (Note 5)	--
(linear olefin) (20)	1	40	4	50	60	68	74	(OAc products)[f]	flash (Note 5)	95
(bicyclic olefin) (100)	5	120	20	250	50	30	76	(OAc product)	60-62 (3)	98
(macrocyclic olefin) (20)	1	40	4	50	40	72	60	(OAc product)	flash (Note 5)	85

113

TABLE (contd.)

[a]Some of the products contain small amounts of the homoallylic isomer (5% or less).

[b]The conversion was determined by NMR.

[c]The yield was not corrected; yield based on consumed starting material is 39%.

[d]Based on consumed olefin the yield is 90%. The starting olefin was a mixture of approximately 62% trans and 32% cis isomer together with 6% cyclododecane. After the reaction about 20% of the starting material could be recovered, now as a mixture of 20% trans, 50% cis olefin and 30% cyclododecane.

[e]This is a GLC yield using n-decane as internal standard.

[f]The product was a 1:1 mixture.

Appendix

Chemical Abstracts Nomenclature (Collective Index Number)

(Registry Number)

2-Cyclohepten-1-yl acetate: 2-Cyclohepten-1-ol, acetate (8,9); (826-13-1)

Cycloheptene (8,9); (628-92-2)

PALLADIUM-CATALYZED COUPLING OF VINYL TRIFLATES WITH
ORGANOSTANNANES: 4-tert-BUTYL-1-VINYLCYCLOHEXENE AND
1-(4-tert-BUTYLCYCLOHEXEN-1-YL)-2-PROPEN-1-ONE
(Cyclohexene, 4-(1,1-dimethylethyl)-1-ethenyl- and
2-Propen-1-one, 1-[4-(1,1-dimethylethyl)-1-cyclohexen-1-yl)

A.

B.

C.

D.

Submitted by William J. Scott,[1a] G. T. Crisp,[1b] and J. K. Stille.[1c]
Checked by Dean R. Lagerwall and Leo A. Paquette.

1. Procedure

Caution! Many organotin compounds are toxic.[2] *Their preparation and use should be carried out in a well-ventilated hood.*

A. *4-tert-Butylcyclohexen-1-yl trifluoromethanesulfonate*. A dry, 2-L, three-necked, round-bottomed flask equipped with a magnetic stirring bar, an argon inlet, and a condenser (Note 1) is charged with 33.0 g (0.214 mol) of 4-tert-butylcyclohexanone (Note 2), 1.5 L of dichloromethane (Note 3), and 49.5 g (0.241 mol) of 2,6-di-tert-butyl-4-methylpyridine (Note 4). The solution is stirred under a static argon atmosphere and cooled to 0°C, at which time the dropwise addition of 40.0 mL (0.238 mol) of trifluoromethanesulfonic anhydride (Note 5) is begun. After the addition is complete, the brown mixture is allowed to warm slowly to room temperature and is stirred at that temperature for 10 hr. At this point, the consumption of starting ketone is verified by thin layer chromatography (hexane; silica gel). If the reaction is incomplete, more trifluoromethanesulfonic anhydride is added and additional time is allowed. The solvent is removed by distillation and the resulting light tan material is treated with 1 L of pentane and heated to reflux for 30 min. The tan salts thus obtained are removed by filtration and washed with five 100-mL portions of pentane (Note 6). The combined pentane solutions are washed with two 250-mL portions each of a 10% hydrochloric acid solution, a 10% sodium hydroxide solution and a saturated sodium chloride solution, dried with magnesium sulfate, filtered through a 6 x 4-cm pad of silica gel (Note 7), and concentrated by distillation. Bulb-to-bulb distillation (Note 8) of the resulting yellow oil at 75-80°C (0.5 mm) gives 43-45 g (70-73%) of the product as a colorless oil (Note 9).

B. *Trimethylvinyltin.* To a dry, 1-L, three-necked, round-bottomed flask, equipped with a Dewar-type condenser cooled to -78°C, a magnetic stirring bar, and a gas inlet leading to a static supply of dry argon (Note 1), are added 11.4 g (0.469 mol) of clean magnesium turnings, 50 mL of dry tetrahydrofuran (Note 10), 3 mL of vinyl bromide, and 0.3 mL of methyl iodide to initiate

117

formation of vinylmagnesium bromide. To this is added a solution of 41 mL (0.624 mol or 66.7 g total) of vinyl bromide in 125 mL of dry tetrahydrofuran via cannula at a rate which maintains a gentle reflux. After addition the mixture is heated to reflux for 1 hr with an oil bath, then cooled to 60°C.

To the resulting slurry of vinylmagnesium bromide (with the condenser still maintained at -78°C) is added via cannula a solution of 61.3 g (0.307 mol) of trimethyltin chloride (Note 11) in 50 mL of dry tetrahydrofuran at a rate suitable to maintain a gentle reflux. The temperature is maintained at 60°C for 5 hr and then the mixture is cooled to room temperature. With the condenser still maintained at -78°C, 200 mL of a saturated ammonium chloride solution is added by syringe at a rate which maintains a gentle reflux, followed by 200 mL of water. The resulting solution is transferred to a separatory funnel with the aid of 200 mL of pentane and the organic layer is washed with 250 mL of a saturated ammonium chloride solution. The combined aqueous layers are back-extracted twice with 250 mL of pentane, and the combined organic layers are washed 5 times with 250 mL of saturated ammonium chloride, 10 times with a 10% hydrochloric acid solution, and twice with a saturated sodium chloride solution, then gravity filtered through a 9 x 4-cm pad of silica gel (Note 7) to give approximately 500 mL of a slightly yellow solution. Pentanes are removed by distillation using a 16-cm Vigreux column and a short-path still-head (Note 12). Continued distillation affords a fraction boiling from 60-90°C, which contains a mixture of 4.9 g of trimethyvinyltin in 16.3 g of tetrahydrofuran (Note 13). At this point the still is cooled to room temperature, the Vigreux column is removed, and the remaining oil is distilled at 95-100°C to give 38-39 g (64-66% purified yield) of trimethylvinyltin (Note 14).

C. 4-tert-Butyl-1-vinylcyclohexene.   A dry, 2-L, one-necked, round-bottomed flask equipped with a magnetic stirring bar, an argon inlet, and a condenser (Note 1) is charged with 1.18 g (1.02 mmol) of tetrakis(triphenyl-phosphine)palladium(0) (Note 15), 12.9 g (0.305 mol) of lithium chloride (Note 16), and 500 mL of tetrahydrofuran (Note 10). This mixture is stirred for 15 min under a static argon atmosphere; then a solution of 28.0 g (0.0979 mol) of 4-tert-butylcyclohexen-1-yl trifluoromethanesulfonate and 19.0 g (0.0997 mol) of trimethylvinyltin in 250 mL of tetrahydrofuran is added, followed by an additional 250 mL of tetrahydrofuran.   The resulting, almost colorless solution is heated to a gentle reflux for 48 hr (Note 17). The mixture is cooled to room temperature and partitioned between 500 mL of water and 250 mL of pentane. The aqueous layer is back-extracted with two 200-mL portions of pentane. The combined organic layers are washed with two 250-mL portions each of a concentrated sodium bicarbonate solution, water, and a concentrated sodium chloride solution, dried over magnesium sulfate, filtered through a 4 x 4-cm pad of silica gel (Note 7) and concentrated by distillation using a 10-cm Vigreux column. Bulb-to-bulb distillation (Note 8) of the resulting yellow oil at 65-68°C (0.55 mm) gives 12.6-12.8 g (78-79%) of the coupled product (Note 18).

D. 1-(4-tert-Butylcyclohexen-1-yl)-2-propen-1-one.   To a dry, 2-L, round-bottomed flask, equipped with a magnetic stirring bar, condenser, and gas inlet connected to a static argon atmosphere (Note 1), are added 1.12 g (0.968 mmol) of tetrakis(triphenylphosphine)palladium(0) (Note 15), 13.2 g (0.312 mol) of lithium chloride (Note 16) and 500 mL of tetrahydrofuran (Note 10), followed by a solution of 28.6 g (0.100 mol) of 4-tert-butylcyclohexen-1-yl trifluoromethansulfonate and 19.1 g (0.100 mol) of trimethylvinyltin in 250 mL of tetrahydrofuran, and then an additional 250 mL of tetrahydrofuran. A gas bag (Note 19) filled with carbon monoxide is attached to the gas inlet and the apparatus is flushed with carbon monoxide. The gas bag is refilled with carbon monoxide and reattached to the gas inlet. The mixture is then heated to 55°C (Note 20). After 2-4 hr, a large amount of the carbon monoxide has been absorbed into solution and the gas bag is refilled and re-attached to the gas inlet. After a total of 40 hr, the reaction mixture darkens and is cooled to room temperature (Note 21). This solution is transferred to a 2-L separatory funnel, diluted with 500 mL of pentane, and washed with two 200-mL portions each of water, saturated sodium bicarbonate solution, water, and saturated sodium chloride solution. The resulting yellow solution is dried over magnesium sulfate, filtered through a 6 x 4-cm pad of silica gel (Note 7), and concentrated using a rotary evaporator. Slow bulb-to-bulb distillation (Notes 8 and 22) of the brown oil at 85-95°C (0.35 mm) gives 14.3-14.5 g (74-75%) of the product as a colorless oil (Note 23).

2. Notes

1. The glassware is dried in an oven at 140°C overnight and assembled warm under a static argon atmosphere.

2. 4-tert-Butylcyclohexanone is purchased from the Aldrich Chemical Company, Inc.

3. Dichloromethane is freshly distilled from calcium hydride.

4. 2,6-Di-tert-butyl-4-methylpyridine can be purchased from the Aldrich Chemical Company, Inc. or prepared from pivaloyl chloride.[3]

5. Trifluoromethanesulfonic anhydride can be purchased from the Aldrich Chemical Company, Inc. or prepared from trifluoromethanesulfonic acid.[4]

6. The solids can range in color from off-white to brown. 2,6-Di-tert-butyl-4-methylpyridine can be recovered by separation of the tan solids between 500 mL of pentane and 650 mL of a 1.2 N sodium hydroxide solution. The aqueous layer is extracted with two 200-mL portions of pentane. The combined organic phases are washed with two 200-mL portions of water and two 200-mL portions of a saturated sodium chloride solution dried with sodium sulfate, filtered through a 6 x 4-cm pad of silica gel (Note 7), and concentrated using a rotary evaporator. Bulb-to-bulb distillation (Note 8) of the yellow oil at 65-75°C (0.55 mm) with trapping of the distillate in two connected receiving flasks cooled to -78°C gives 42.1-48.1 g (85-97%) of colorless oil, which solidifies on standing.

7. Woelm 230-400 mesh silica gel is used.

8. An Aldrich Chemical Company, Inc. Kugelrohr apparatus is used.

9. 4-tert-Butylcyclohexen-1-yl trifluoromethanesulfonate has the following properties: mp 17°C; TLC R_f (hexanes): 0.65; IR (neat) cm^{-1}: 1698, 1440, 1425, 1250, 1225; ^1H NMR (CDCl$_3$, 270 MHz) δ: 0.89 (s, 9 H), 1.26-1.38 (m, 3 H), 1.90-2.18 (m, 2 H), 2.23-2.39 (m, 2 H), 5.73-5.75 (m, 1 H): ^{13}C NMR (CDCl$_3$, 68 MHz) δ: 23.9, 25.4, 27.2 (3C), 28.6, 32.0, 43.1, 118.3, 118.7 (q, J = 319, CF$_3$), 149.3.

10. Tetrahydrofuran is freshly doubly-distilled from potassium.

11. Trimethyltin chloride can be purchased from Strem Chemicals, Inc. or prepared by the reaction of tetramethyltin with tin tetrachloride as follows: To a 100-mL, round-bottomed flask, equipped with a magnetic stirring bar, and a septum, and a gas inlet connected to a static argon atmosphere, containing 41.2 g (0.230 mol) of tetramethyltin cooled to -20°C with a dry ice/carbon tetrachloride slurry, is added 9.0 mL (0.0769 mol) of tin tetrachloride at a slow dropwise rate. After the addition is complete, the

121

mixture is heated to 60°C for 16 hr. The mixture is cooled to room temperature to afford 61.3 g (100% yield) of a colorless solid.

12. In order to maximize the yield of trimethylvinyltin, pentanes should be removed as carefully as possible. The authors employed only 14/20-ground glassware in the distillation and carefully controlled the oil bath temperature to maintain a collection rate of approximately 0.3 mL per min.

13. The ratio of trimethylvinyltin to tetrahydrofuran is determined by NMR. The trimethylvinyltin/THF solution may be used in palladium-catalyzed coupling reactions without further purification.

14. Trimethylvinyltin is >97% pure as shown by gas chromatography on a 1/8" x 6' column packed with 6% SP-2100 on Supelcoport, 80-100 mesh, operated at 50°C. The relative retention times are: 1.9 min for tetrahydrofuran, 4.0 min for trimethylvinyltin, and 7.0 min for trimethyltin chloride (not seen). The distilled product has the following properties: ^{1}H NMR (CDCl$_3$, 270 MHz) δ: 0.12 (s, 9 H), 5.66 (dd, 1 H, J = 3.4, 20.5), 6.11 (dd, 1 H, J = 3.6, 13.9), 6.52 (dd, 1 H, J = 13.9, 20.5); ^{13}C NMR (CDCl$_3$, 68 MHz) δ: -9.9 (3C) 133.2, 140.0.

15. Tetrakis(triphenylphosphine)palladium(0) can be purchased from Strem Chemicals, Inc. or prepared from palladium chloride.[5] On standing for a period of time (> a few weeks) the catalyst gradually darkens, turning tan in the absence of oxygen or turning green in the presence of oxygen. However, the coupling reactions run equally well with catalyst that has aged for a year.

16. Lithium chloride is dried at 140°C for 24 hr prior to use.

17. The progress of the reaction is conveniently monitored by gas chromatography using a 1/8" x 6' column packed with 6% SP-2100 on Supelcoport, 80-100 mesh, operated at 50°C for 4 min, then heated at 15°C/min to 250°C.

The relative retention times are: 4.0 min for trimethylvinyltin, 6.5 min for trimethyltin chloride, 12.0 min for 4-tert-butyl-1-vinylcyclohexene, and 13.1 min for 4-tert-butylcyclohexen-1-yl trifluoromethanesulfonate. Because of the extreme volatility of trimethylvinyltin, it may be necessary to add additional small amounts in order to drive the reaction to completion.

18. 4-tert-Butyl-1-vinylcyclohexene has the following properties: bp 45°C (0.1 mm); TLC R_f (hexanes): 0.74; IR (neat) cm^{-1}: 3100, 3020, 1650, 1610, 1395, 1365, 985, 890; ^{1}H NMR (CDCl$_3$, 270 MHz) δ: 0.87 (s, 9 H), 1.08-1.34 (m, 3 H), 1.84-2.36 (m, 4 H), 4.88 (d, 1 H, J = 10.7) 5.04 (d, 1 H, J = 17.5), 5.73-5.75 (m, 1 H), 6.35 (dd, 1 H, J = 10.7, 17.5); ^{13}C NMR (CDCl$_3$, 68 MHz) δ: 23.8, 25.3, 27.2 (3C), 27.4, 32.2, 44.4, 109.7, 129.8, 136.0, 139.7.

19. The gas bag can be purchased from the Fisher Scientific Company and is filled to approximately 5 psig with carbon monoxide.

20. Refluxing conditions must be avoided in order to maximize the amount of carbon monoxide in solution.

21. The progress of the reaction is conveniently monitored by gas chromatography on a 1/8" x 6' column packed with 6% SP-2100 on Supelcoport, 80-100 mesh, operated at 50°C for 4 min, then heated at 15°C/min to 250°C. The relative retention times are: 4.0 min for trimethylvinyltin, 6.5 min for trimethyltin chloride, 13.1 min for 4-tert-butylcyclohexen-1-yl trifluoromethanesulfonate, and 14.7 min for 1-(4-tert-butylcyclohexen-1-yl)-2-propen-1-one. Because of the extreme volatility of trimethylvinyltin, it may be necessary to add additional small amounts of this reagent in order to drive the reaction to completion.

22. The purification procedure occasionally leads to product contaminated with organotins. The submitters have found that careful washing with water minimizes this problem. The checkers found that distillation of product at a slow rate allows the unwanted tin to escape to the cold trap.

23.   1-(4-tert-Butylcyclohexen-1-yl)-2-propen-1-one has the following properties: bp 75°C (0.1 mm); IR (neat) cm^{-1}:   1665, 1645, 1612; ^{1}H NMR (CDCl$_3$, 270 MHz) δ: 0.81 (s, 9 H), 1.21-2.65 (m, 7 H), 5.58 (d, 1 H, J = 9.0), 6.14 (d, 1 H, J = 17.2), 6.75-7.00 (m, 2 H); ^{13}C NMR (CDCl$_3$, 68 MHz) δ: 23.3, 24.6, 26.9 (3C), 27.8, 32.0, 43.4, 127.1, 131.5, 139.4, 141.1, 190.8.

3. Discussion

The conversion of a ketone into a substituted olefin is classically achieved by the addition of a Grignard reagent to a ketone, followed by the dehydration of the resulting alcohol.   Such a scheme can often lead to a mixture of regioisomers.   By converting the ketone into a vinyl iodide,[6] followed by a cuprate coupling reaction,[7] it is possible to form selectively the less-hindered, substituted olefin. Group 10[8]-catalyzed coupling reactions of vinyl iodides also lead to the formation of olefins in good yields.[9-11] However, the synthesis of the more hindered vinyl iodides can be quite difficult.

A number of enolate derivatives have recently been offered as alternatives to vinyl iodides.   The advantage to such a scheme lies in the ability to form regioselectively either the kinetic or the thermodynamic enolate using known methodology,[12] and then to trap that enolate to give the required derivative.   In general, enolate derivatives, such as methyl vinyl ethers,[13] silyl enol ethers,[14] and enol phosphates,[15] have undergone coupling only in the presence of nickel catalysts, thus requiring the use of very strong nucleophiles.   However, such nucleophiles severely restrict the functionality which may be present in either the enolate derivative or the nucleophile.

Vinyl trifluoromethanesulfonates[16] have provided a solution to this problem. Vinyl trifluoromethanesulfonates can be formed by the action of trifluoromethanesulfonic anhydride with a ketone.[4,16] Enolates may be trapped using N-phenyltrifluoromethanesulfonimide to form selectively either the kinetic or thermodynamic derivative.[17-19] The resulting enolate derivatives couple readily with organocuprates.[20] Palladium-catalyzed coupling reactions may also be run to give directly coupled products,[19] Heck-type coupled products,[21-23] or reduced products.[19-24] Direct coupling reactions of vinyl trifluoromethanesulfonates have been used in the synthesis of pleraplysillin-1,[19] the synthesis of cardenolides,[22] the synthesis of vinylstannanes,[25-27] and in intramolecular cyclization reactions.[28-30]

The synthesis of ketones is very important to the organic chemist. Two common methods involve the addition of Grignard reagents to aldehydes, followed by oxidation of the secondary alcohol, and the addition of organolithium reagents to carboxylic acids.[31] In addition, acid chlorides have been coupled with Grignard reagents,[32,33] organoaluminum reagents,[34] organocadmium reagents,[33] organocuprates,[7] or organozinc reagents[33] to give the corresponding ketone. More recently, the palladium-catalyzed coupling of acid chlorides with organozinc reagents,[35] organostannanes,[36] or organomercury reagents[37] has provided a very mild method for ketone synthesis.

In order to avoid the necessity of using acid chlorides in the coupling reactions, the palladium-catalyzed coupling of electrophiles in the presence of carbon monoxide was developed.[38,39] Again, the necessity of using vinyl iodides limits this methodology. Upon palladium-catalyzed coupling of vinyl trifluoromethanesulfonates in the presence of lithium chloride, the desired enones are formed in good yield.[40,41] The carbonylative coupling reaction has been used in the synthesis of $(\pm)-\Delta^{9,12}$-capnellene.[40]

1. a) Department of Chemistry, University of Iowa, Iowa City, IA 52242; b) Department of Organic Chemistry, University of Melbourne, Parkville, Victoria, 3052, Australia; c) Department of Chemistry, Colorado State University, Fort Collins, CO 80523.

2. Krigman, M. R.; Silverman, A. P. *Neurotoxicology* **1984**, *5*, 129-139.

3. Anderson, A. G.; Stang, P. J. *Org. Synth.* **1981**, *60*, 34-40.

4. Stang, P. J.; Dueber, T. E. *Org. Synth., Collect. Vol. VI* **1988**, 757-761.

5. Coulson, D. R. *Inorg. Synth.* **1972**, *13*, 121-124.

6. Barton, D. H. R.; Bashiardes, G.; Fourrey, J.-L. *Tetrahedron Lett.* **1983**, *24*, 1605-1608.

7. Posner, G. H. *Org. React.* **1975**, *22*, 253-400.

8. The group notation has been changed in accord with recent recommendations by the IUPAC and ACS nomenclature committees. Group I became groups 1 and 11, Group II became groups 2 and 12, Group III became groups 3 and 13, etc.

9. Heck, R. F. *Acc. Chem. Res.* **1979**, *12*, 146-151.

10. Negishi, E.-i. *Acc. Chem. Res.* **1982**, *15*, 340-348.

11. Stille, J. K.; Groh, B. L. *J. Am. Chem. Soc.* **1987**, *109*, 813-817.

12. d'Angelo, J. *Tetrahedron* **1976**, *32*, 2979-2990.

13. Wenkert, E.; Michelotti, E. L.; Swindell, C. S.; Tingoli, M. *J. Org. Chem.* **1984**, *49*, 4894-4899.

14. Hayashi, T.; Katsuro, Y.; Kumada, M. *Tetrahedron Lett.* **1980**, *21*, 3915-3918.

15. Takai, K.; Sato, M.; Oshima, K.; Nozaki, H. *Bull. Chem. Soc. Jpn.* **1984**, *57*, 108-115.

16. For a review on the chemistry of vinyl trifluoromethanesulfonates, see: Stang, P. J.; Hanack, M.; Subramaniam, L. R.; *Synthesis*, **1982**, 85-126. For a review of the use of vinyl trifluoromethanesulfonates as electrophiles in carbon-carbon bond forming reactions, see: Scott, W. J.; McMurry, J. E. *Acc. Chem. Res.* **1988**, *21*, 47-54.

17. McMurry, J. E.; Scott, W. J. *Tetrahedron Lett.* **1983**, *24*, 979-982.

18. Crisp, G. T.; Scott, W. J. *Synthesis* **1985**, 335-337.

19. Scott, W. J.; Crisp, G. T.; Stille, J. K. *J. Am. Chem. Soc.* **1984**, *106*, 4630-4632; Scott, W. J.; Stille, J. K. *J. Am. Chem. Soc.* **1986**, *108*, 3033-3040.

20. McMurry, J. E.; Scott, W. J. *Tetrahedron Lett.* **1980**, *21*, 4313-4316.

21. Cacchi, S.; Morera, E.; Ortar, G. *Tetrahedron Lett.* **1984**, *25*, 2271-2274.

22. Harnisch, W.; Morera, E.; Ortar, G. *J. Org. Chem.* **1985**, *50*, 1990-1992.

23. Scott, W. J.; Peña, M. R.; Swärd, K.; Stoessel, S. J.; Stille, J. K. *J. Org. Chem.* **1985**, *50*, 2302-2308.

24. Cacchi, S.; Morera, E.; Ortar, G. *Tetrahedron Lett.* **1984**, *25*, 4821-4824.

25. Piers, E.; Tse, H. L. A. *Tetrahedron Lett.* **1984**, *25*, 3155-3158.

26. Matsubara, S.; Hibino, J.; Morizawa, Y.; Oshima, K.; Nozaki, H. *J. Organomet. Chem.* **1985**, *285*, 163-172.

27. Wulff, W. D.; Peterson, G. A.; Bauta, W. E.; Chan, K.-S.; Faron, K. L.; Gilbertson, S. R.; Kaesler, R. W.; Yang, D. C.; Murray, C. K. *J. Org. Chem.* **1986**, *51*, 277-279.

28. Piers, E.; Friesen, R. W.; Keay, B. A. *J. Chem. Soc., Chem. Commun.* **1985**, 809-810.

29. Piers, E.; Friesen, R. W. *J. Org. Chem.* **1986**, *51*, 3405-3406.

30. Stille, J. K.; Tanaka, M. *J. Am. Chem. Soc.* **1987**, *109*, 3785-3786.

31. Jorgenson, M. J. *Org. React.* **1970**, *18*, 1-97.

32. Sato, F.; Inoue, M. Oguro, K.; Sato, M. *Tetrahedron Lett.* **1979**, *20*, 4303-4306.

33. Shirley, D. A. *Org. React.* **1954**, *8*, 28-58.

34. Reinheckel, H.; Haage, K.; Ludwig, H. *J. Prakt. Chem.* **1975**, *317*, 359-368.

35. Negishi, E.-i.; Bagheri, V.; Chatterjee, S.; Luo, F.-T.; Miller, J. A.; Stoll, A. T. *Tetrahedron Lett.* **1983**, *24*, 5181-5184.

36. Milstein, D.; Stille, J. K. *J. Org. Chem.* **1979**, *44*, 1613-1618.

37. Takagi, K.; Okamoto, T.; Sakakibara, Y.; Ohno, A.; Oka, S.; Hayama, N. *Chem. Lett.* **1975**, 951-954.

38. Beletskaya, I. P. *J. Organometal. Chem.* **1983**, *250*, 551-564.

39. Stille, J. K. *Pure Appl. Chem.* **1985**, *57*, 1771-1780; Stille, J. K. *Angew. Chem., Intern. Ed. Engl.* **1986**, *25*, 508-524.

40. Crisp, G. T.; Scott, W. J.; Stille, J. K. *J. Am. Chem. Soc.* **1984**, *106*, 7500-7506.

41. Baillargeon, V. P.; Stille, J. K. *J. Am. Chem. Soc.* **1986**, *108*, 452-461.

Appendix
Chemical Abstracts Nomenclature (Collective Index Number)
(Registry Number)

4-tert-Butyl-1-vinylcyclohexene: Cyclohexene, 4-tert-butyl-1-vinyl- (8); Cyclohexene, 4-(1,1-dimethylethyl)-1-ethenyl- (9); (33800-81-6)

1-(4-tert-Butylcyclohexen-1-yl)-2-propen-1-one: 2-Propen-1-one, 1-[4-(1,1-dimethylethyl)-1-cyclohexen-1-yl]- (11); (92622-56-5)

4-tert-Butylcyclohexen-1-yl trifluoromethanesulfonate: Methanesulfonic acid, trifluoro-, 4-(1,1-dimethylethyl)-1-cyclohexen-1-yl ester (10); (77412-96-5)

4-tert-Butylcyclohexanone: Cyclohexanone, 4-tert-butyl- (8); Cyclohexanone,
4-(1,1-dimethylethyl)- (9); (98-53-3)

2,6-Di-tert-butyl-4-methylpyridine: Pyridine, 2,6-bis(1,1-dimethylethyl)-4-
methyl- (9); (38222-83-2)

Trifluoromethanesulfonic anhydride: Methanesulfonic acid, trifluoro-,
anhydride (8,9); (358-23-6)

Trimethylvinyltin: Stannane, trimethylvinyl- (8); Stannane,
ethenyltrimethyl- (9); (754-06-3)

Trimethyltin chloride: Stannane, chlorotrimethyl- (8,9); (1066-45-1)

Tetrakis(triphenylphosphine)palladium(0): Palladium,
tetrakis(triphenylphosphine)- (8); Palladium, tetrakis(triphenylphosphine)-,
(T-4)- (9); (14221-01-3)

A. $C_4H_9C\equiv CH$ + $\longrightarrow$

B.   + $\xrightarrow[\text{NaOEt}]{\text{PdCl}_2(\text{PPh}_3)_2}$

Submitted by Norio Miyaura and Akira Suzuki.[1]

Checked by Albert L. Casalnuovo, Thomas S. Kline, Jr., and Bruce E. Smart.

1. Procedure

A. *(E)-1-Hexenyl-1,3,2-benzodioxaborole.* A 25-mL, three-necked, round-bottomed flask is equipped with a magnetic stirring bar, thermometer, rubber septum, 10-mL addition funnel, and a reflux condenser. The apparatus is connected through the condenser to a nitrogen source and an oil bubbler (Note 1). The flask is charged with 4.9 g (60 mmol) of 1-hexyne (Note 2) through the addition funnel. While the solution is stirred slowly, 6.7 mL (60 mmol) of catecholborane (Note 3) is injected by syringe through the septum cap. The exothermic reaction is maintained at 60-70°C by intermittent cooling in an ice-water bath. The reaction mixture is allowed to cool to room temperature and is stirred for 15 min. The rubber septum is replaced by a glass stopper,

130

and the mixture is heated to 60°C and stirred for an additional 2 hr. The flask is cooled to room temperature, the condenser is replaced by a short-path distillation head, and the mixture is distilled at reduced pressure to give 9.5-10.5 g (78-87%) of clear, colorless product, bp 75-76°C (0.10 mm) [lit.[2] bp 82°C (0.25 mm)] (Note 4).

B. *(1Z,3E)-1-Phenyl-1,3-octadiene.* A 500-mL, three-necked, round-bottomed flask equipped with a magnetic stirring bar, a reflux condenser to which a nitrogen inlet tube and oil bubbler are attached, a glass stopper, and an addition funnel is flushed with nitrogen and charged with 9.5 g (47 mmol) of (E)-1-hexenyl-1,3,2-benzodioxaborole and 200 mL of benzene (Note 5). The solution is stirred and 8.4 g (46 mmol) of (Z)-β-bromostyrene (Note 6), 50 mL of 2 M sodium ethoxide in ethanol (Note 7), and finally 0.28 g (0.4 mmol) of dichlorobis(triphenylphosphine)palladium(II) (Note 8) are added. The mixture is refluxed for 3 hr. The light brown solution containing a white precipitate of sodium bromide is cooled to room temperature, treated with 60 mL of 3 M sodium hydroxide and 6 mL of 30% hydrogen peroxide, and stirred at room temperature for 1 hr (Note 9). The organic layer is separated, washed four times with 50 mL of 3 M sodium hydroxide (Note 10), and dried over anhydrous magnesium sulfate. The drying agent is removed by filtration and the filtrate is concentrated on a rotary evaporator. The residual oil is distilled under reduced pressure (Note 11) to give 7.0 g (82%) of (1Z,3E)-1-phenyl-1,3-octadiene as a clear, colorless liquid, bp 80°C (0.15 mm) (Note 12).

2. Notes

1. All glassware was pre-dried in an oven at 130°C for 3 hr, assembled while hot, and allowed to cool under a stream of nitrogen.

2. The preparation of 1-hexyne is described in *Org. Synth.*, *Collect Vol.* IV **1963**, 117. The checkers obtained 1-hexyne from Aldrich Chemical Company, Inc., and distilled it prior to use.

3. Catecholborane (1,3,2-benzodioxaborole) with a purity of 95% was purchased from Aldrich Chemical Company, Inc. and purified by distillation under nitrogen, bp 58°C (52 mm). For the distillation and handling of air and moisture sensitive compounds, see references 3-5. Catecholborane is a liquid at room temperature, and the neat material is 9.0 M in catecholborane.[3] The preparation of catecholborane from borane and catechol has been reported.[3]

4. The submitters report bp 86-87°C (0.3 mm). (E)-1-Hexenyl-1,3,2-benzodioxaborole is quite air stable, but it can slowly hydrolyze to boronic acid and turn brown on repeated use in air. The submitters recommend storing it at refrigerator temperature in a bottle purged with nitrogen and capped with a rubber septum. Alternatively, the crude 1-hexenyl-1,3,2-benzodioxaborole can be used for the next coupling reaction without purification. In this case, the unreacted 1-hexyne should be removed under reduced pressure (0.1 mm for 30 min), because it also reacts with (Z)-β-bromostryrene to afford (1Z)-1-phenyl-1-octen-3-yne. In this manner the expected diene was obtained in a yield of 86%.

5. Benzene was obtained from Fisher Scientific Company and redistilled before use.

6. (Z)-β-Bromostyrene was prepared by the procedure described in *Org. Synth.* **1984**, *62*, 39.

7. The sodium ethoxide solution was prepared by dissolving 2.3 g of sodium in 50 mL of anhydrous ethanol and was used immediately.

8. The palladium catalyst is prepared as follows. A 100-mL, one-necked, round-bottomed flask equipped with a magnetic stirring bar and a reflux

condenser connected to a nitrogen inlet is flushed with nitrogen and charged with 1.00 g (5.64 mmol) of palladium chloride (Johnson Matthey, Inc.), 3.25 g (12.4 mmol) of triphenylphosphine (Aldrich Chemical Company, Inc.), and 30 mL of benzonitrile (Aldrich Chemical Company, Inc.). The mixture is stirred, gradually heated to 180°C, and held at that temperature for 20 min. The clear, red solution that results is allowed to cool slowly to room temperature and stand overnight. The bright yellow crystals which precipitate are collected by filtration, washed with three 10-mL portions of ether, and dried at reduced pressure to give 3.90 g (5.55 mmol) of dichlorobis(triphenyl-phosphine)palladium(II).

9. This operation removes most of the palladium-containing compounds. Any unreacted 1-hexenylboronate is oxidized to hexenal.

10. Catechol must be washed out completely because it is difficult to remove by distillation. A solution of catechol in aqueous sodium hydroxide turns dark brown on treatment with hydrogen peroxide or on standing in air.

11. The residual oil can be purified before distillation by filtering it through a short (20 cm) silica gel column (70-230 mesh) using hexanes as an eluent. This effectively removes traces of catechol and palladium-containing compounds.

12. Gas chromatographic analysis of the product (Hewlett-Packard fused silica, cross-linked methylsilicone capillary column, 25 m x 20 mm, column temperature 100-270°C, injection temperature 250°C) shows that the product is over 99% chemically and isomerically pure. (Z,E)-1-Phenyl-1,3-octadiene shows the following spectral properties: IR (neat) cm^{-1}: 1640, 1595, 1490, 985; ^1H NMR ($CDCl_3$) δ: 0.89 (t, 3 H, J = 7.1), 1.25-1.45 (m, 4 H), 2.05-2.20 (m, 2 H), 5.87 (d of t, 1 H, J = 7.1, 15, PhC=C-C=CH), 6.21 (d of d, 1 H, J = 11.2, 11.6, PhC=CH), 6.30 (d, 1 H, J = 11.6, PhCH=C), 6.60 (d of d, 1 H, J = 11.2, 15, PhC=C-CH=C), and 7.15-7.40 (m 5 H, aromatic).

3. Discussion

The procedure described here is an example of a general method for preparing conjugated alkadienes by the palladium-catalyzed reaction of 1-alkenylboranes or boronates with vinylic halides. Hydroboration of 1-alkynes with catecholborane is a standard method for obtaining (E)-1-alkenylboronates (1).[2,3] Several different types of alkenylboranes and boronates (2-4) are now available as reagents for the cross-coupling reaction with vinyl halides.

$$1 \qquad 2^{6,7} \qquad 3^{8} \qquad 4^{9}$$

These alkenylboron derivatives react not only with 1-alkenyl halides but also with a variety of other organic halides, including 1-bromo-1-alkynes,[6] aryl haldies,[7,9,10] and allylic or benzylic halides,[11] in the presence of a palladium catalyst and base. Both $Pd(PPh_3)_4$ and $PdCl_2(PPh_3)_2$ are excellent catalysts for most of the reactions. A base is generally required for successful coupling. Sodium ethoxide (2 equiv) in ethanol-benzene, which is used in the procedure described here, gives high yields with most 1-bromo-1-alkenes. For 1-iodo-1-alkenes, aqueous sodium hydroxide in tetrahydrofuran[12] or aqueous 4 M potassium hydroxide (3 equiv) in benzene[9] can give better results. Alkoxides and hydroxides normally accelerate the reaction, but the choice of base depends upon its compatibility with the particular organic halide. For the coupling reaction with 3-halo-2-cyclohexen-1-one[13] a relatively weak base, such as sodium acetate in methanol, works well. The reaction with 1-bromo-2-phenylthio-1-alkenes[14] is successfully carried out using aqueous potassium hydroxide. For the reaction of (E)-2-

ethoxyvinylborane with aryl halides,[15] a suspension of sodium hydroxide in tetrahydrofuran gives better results than homogeneous base. The versatility of these methods has been reviewed.[8,16]

In addition to alkenylboron compounds, alkenylalane,[17] alkenylzirconium,[18] alkenyltin,[19] alkenylcopper,[20] and alkenylmagnesium[21] reagents are reported to undergo a related alkenyl-alkenyl coupling reaction to give 1,3-alkadienes.

1. Department of Applied Chemistry, Faculty of Engineering, Hokkaido University, Sapporo 060, Japan.

2. Brown, H. C.; Gupta, S. K. *J. Am. Chem. Soc.* **1972**, *94*, 4370.

3. Brown, H. C.; Kramer, G. W.; Levy, A. B.; Midland, M.M. "Organic Synthesis via Boranes", Wiley: New York, 1975; pp. 63-65, 226-231, and 241-245.

4. Gill, G. B.; Whiting, D. A. *Aldrichmica Acta* **1986**, *19*, 31.

5. Lane, C. F.; Kramer, G. W. *Aldrichmica Acta* **1977**, *10*, 11.

6. Miyaura, N.; Yamada, K.; Suginome, H.; Suzuki, A. *J. Am. Chem. Soc.* **1985**, *107*, 972.

7. Miyaura, N.; Satoh, M.; Suzuki, A. *Tetrahedron Lett.* **1986**, *27*, 3745.

8. Suzuki, A. *Pure and Appl. Chem.* **1986**, *58*, 629.

9. Satoh, M.; Miyaura, N.; Suzuki, A. *Chem. Lett.* **1986**, 1329.

10. Miyaura, N.; Suzuki, A. *J. Chem. Soc., Chem. Commun.* **1979**, 866.

11. Miyaura, N.; Yano, T.; Suzuki, A. *Tetrahedron Lett.* **1980**, *21*, 2865.

12. Cassani, G.; Massardo, P.; Piccardi, P. *Tetrahedron Lett.* **1983**, *24*, 2513.

13. Satoh, N.; Ishiyama, T.; Miyaura, N.; Suzuki, A. *Bull. Chem. Soc. Jpn.* **1987**, *60*, 3471.

14. Ishiyama, T.; Miyaura, N.; Suzuki, A. *Chem. Lett.* **1987**, 25.

15. Miyaura, N.; Maeda, K.; Suginome, H.; Suzuki, A. *J. Org. Chem.* **1982**, *47*, 2117.

16. Suzuki, A. *Pure and Appl. Chem.* **1985**, *57*, 1749.

17. (a) Baba, S.; Negishi, E.-i. *J. Am. Chem. Soc.* **1976**, *98*, 6729; (b) Negishi, E.-i.; Okukado, N.; King, A. O.; Van Horn, D. E.; Spiegel, B. I. *J. Am. Chem. Soc.* **1978**, *100*, 2254; (c) Negishi, E.-i. *Acc. Chem. Res.* **1982**, *15*, 340.

18. (a) Okukado, N.; Van Horn, D. E.; Klima, W. L.; Negishi, E.-i. *Tetrahedron Lett.* **1978**, 1027; (b) Negishi, E.-i.; Takahashi, T. *Aldrichmica Acta* **1985**, *18*, 31.

19. Stille, J. K.; Groh, B. L. *J. Am. Chem. Soc.* **1987**, *109*, 813.

20. (a) Jabri, N.; Alexakis, A.; Normant, J. F. *Tetrahedron Lett.* **1981**, *22*, 959; (b) Jabri, N.; Alexakis, A.; Normant, J. F. *Tetrahedron Lett.* **1982**, *23*, 1589.

21. Dang, H. P.; Linstrumelle, G. *Tetrahedron Lett.* **1978**, 191.

Appendix

Chemical Abstracts Nomenclature (Collective Index Number);

(Registry Number)

(1Z,3E)-1-Phenyl-1,3-octadiene: Benzene, 1,3-octadienyl-, (Z,E)- (9);

(39491-66-2)

(E)-1-Hexenyl-1,3,2-benzodioxaborole: 1,3,2-Benzodioxaborole, 2-(1-hexenyl)-,

(E)- (9); (37490-22-5)

1-Hexyne (8,9); (693-02-7)

Catecholborane: 1,3,2-Benzodioxaborole (8,9); (274-07-7)

(Z)-β-Bromostyrene: Styrene, β-bromo-, (Z)- (8); Benzene,

(2-bromoethenyl)- (Z)- (9); 588-73-8)

Dichlorobis(triphenylphosphine)palladium(II): Palladium,

dichlorobis(triphenylphosphine)- (8,9); (13965-03-2)

Palladium chloride (8,9); (7647-10-1)

Triphenylphosphine: Phosphine, triphenyl- (8,9); (603-35-0)

PALLADIUM-CATALYZED REDUCTION OF VINYL TRIFLUOROMETHANESULFONATES

TO ALKENES: CHOLESTA-3,5-DIENE

A.

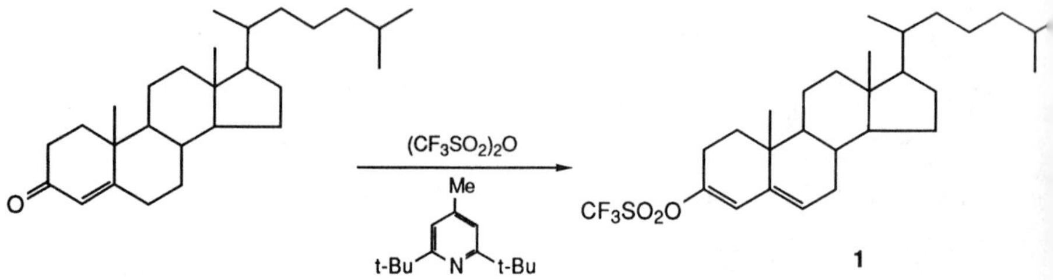

$(CF_3SO_2)_2O$

Me

t-Bu—N—t-Bu

CF_3SO_2O

1

B. 1

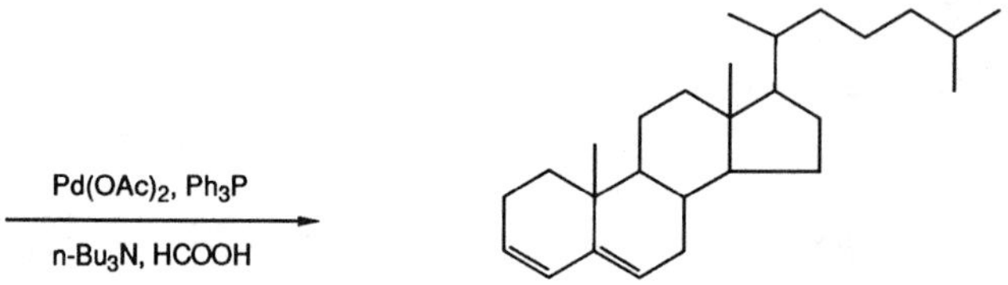

$Pd(OAc)_2$, Ph_3P

n-Bu_3N, HCOOH

Submitted by Sandro Cacchi,[1] Enrico Morera,[2] and Giorgio Ortar.[2]

Checked by Sean Kerwin, Christopher Schmid, and Clayton H. Heathcock.

1. Procedure

A. Cholesta-3,5-dien-3-yl trifluoromethanesulfonate. A dry, 250-mL, two-necked, round-bottomed flask, equipped with a magnetic stirring bar, rubber septum, and pressure-equalizing 100-mL dropping funnel fitted with a calcium chloride drying tube is charged with 4.62 g (22.5 mmol) of 2,6-di-tert-butyl-4-methylpyridine (Note 1) and 60 mL of dry dichloromethane (Note 2). Then 3.08 mL (18.75 mmol) of trifluoromethanesulfonic anhydride (Note 3) is added rapidly from a syringe and 5.77 g (15 mmol) of cholest-4-en-3-one (Note 4) diluted in 40 mL of dry dichloromethane is added through the dropping funnel, dropwise and with stirring, during 15-20 min. The mixture is stirred for an additional 1 hr at room temperature. During this period the solution turns slightly pink and a white precipitate separates. The solvent is removed with a rotary evaporator and the residue is combined with 100 mL of diethyl ether. The white pyridinium trifluoromethanesulfonate salt is filtered off and washed with additional diethyl ether (3 x 50 mL). The ethereal solution is washed with cold 2 N hydrochloric acid (2 x 100 mL) and with saturated sodium chloride solution (3 x 100 mL), dried over anhydrous potassium carbonate, and concentrated at reduced pressure. The solid residue (7.21-7.40 g) is recrystallized from hexane to give 6.46-6.72 g (83-87%) of cholesta-3,5-dien-3-yl trifluoromethanesulfonate as white crystals (Note 5), mp 125-126°C (Note 6).

B. *Cholesta-3,5-diene*. A 50-mL, two-necked, round-bottomed flask, equipped with a magnetic stirring bar and a reflux condenser with a nitrogen inlet at the top is charged with 5.00 g (9.68 mmol) of cholesta-3,5-dien-3-yl trifluoromethanesulfonate (1), 6.92 mL (29.03 mmol) of tributylamine (Note 7), 0.043 g (0.19 mmol) of palladium acetate, 0.100 g (0.38 mmol) of triphenylphosphine, and 20.2 mL of N,N-dimethylformamide. The mixture is gently flushed with nitrogen for 1-2 min and capped with a rubber septum. Formic acid, 99%, 0.73 mL (19.42 mmol) is added from a syringe dropwise and with swirling during 2-3 min. The resulting mixture is warmed in an oil bath at 60°C for 1 hr with continuous stirring under nitrogen. During this period the mixture becomes black. The contents of the flask are poured into 50 mL of 2 N hydrochloric acid and extracted with two 75-mL portions of ethyl ether. The combined organic phases are then washed with 50 mL of 2 N hydrochloric acid, 15 mL of saturated sodium bicarbonate solution, two 10-mL portions of saturated sodium chloride solution, and dried over anhydrous magnesium sulfate. The drying agent is removed by filtration, the ether is evaporated at reduced pressure, and the solid residue (3.92-4.16 g) is purified by open-column chromatography on 100 g of basic aluminum oxide (Note 8) using hexane as eluent to give 3.12-3.22 g of nearly pure cholesta-3,5-diene which is recrystallized from acetone to give a first crop (2.92-3.00 g) as white needles (Note 9), mp 81.5-82.5°C [lit.[3] mp 79.5-80°C] (Note 6) and a second crop (0.11-0.15 g, 85-88% overall yield), mp 79.5-80.5°C (Note 6).

2. Notes

1. A commercial sample of 2,6-di-tert-butyl-4-methylpyridine from Fluka AG was purified through a short column of silica gel by eluting with hexane. Alternatively it may be prepared according to the procedure reported in Organic Syntheses.[4]

2. Reagent grade dichloromethane is dried by passing over a column of aluminum oxide (activity I).

3. Trifluoromethanesulfonic anhydride from Fluka AG was stirred over phosphorus pentoxide for 18 hr and distilled. It can also be prepared from trifluoromethanesulfonic acid (Fluka AG) according to the procedure described in Organic Syntheses.[5]

4. Cholest-4-en-3-one was purchased from Fluka AG and used without further purification.

5. Spectral data are as follows: ^{1}H NMR (90 MHz, $CDCl_3$) δ: 0.69 (s, 3 H, 13-CH_3), 0.82 (s, 3 H, 10-CH_3), 5.62 (m, 1 H, C-6 H), 6.02 (m, 1 H, C-4 H); MS m/e: 516 (M^+).

6. Melting points are uncorrected and were determined with a Köfler hot-stage apparatus.

7.   Tributylamine, palladium acetate, triphenylphosphine from Fluka AG and N,N-dimethylformamide and formic acid from Farmitalia Carlo Erba Chemicals were used without further purification.

8.   Basic aluminum oxide (activity I) is available from Merck & Company, Inc.

9.   This compound has the following physical properties: ^{1}H NMR (90 MHz, $CDCl_3$) δ: 0.69 (s, 3 H, 13-CH_3), 0.82 (s, 3 H, 10-CH_3), 5.4 (m, 1 H, C-6 H), 5.59 (m, 1 H, C-3 H), 5.71 (d, 1 H, J = 10, C-4 H); $[\alpha]_D$ ($CHCl_3$, 1%) -115° (lit.[3] $[\alpha]_D$ -123°).

3. Discussion

The present preparation illustrates a general and convenient method for a two-step deoxygenation of carbonyl compounds to olefins.[6] Related procedures comprise the basic decomposition of p-toluenesulfonylhydrazones,[7] the hydride reduction of enol ethers,[8] enol acetates,[9] enamines,[10] the reduction of enol phosphates (and/or enol phosphorodiamidates) by lithium metal in ethylamine (or liquid ammonia),[11] the reduction of enol phosphates by titanium metal under aprotic conditions,[12] the reduction of thioketals by Raney nickel,[13] and the reduction of vinyl sulfides by Raney nickel in the presence of isopropylmagnesium bromide.[14]

Following our first report on the palladium-catalyzed reaction of vinyl triflates with olefins[15a] (Heck-type reaction), oxidative insertion of palladium(0) into the carbon-oxygen bond of easily available vinyl triflates[16] has proved to be a general method for the generation of σ-vinyl palladium intermediates which can react directly with a variety of olefinic systems,[15] carbon monoxide and alcohols or amines,[17] or 1-alkynes,[18] to give conjugated dienes, α,β-unsaturated esters or amides, or conjugated enynes, respectively. Palladium-catalyzed coupling of vinyl triflates with organostannanes has also been reported.[19]

σ-Vinyl palladium triflates are smoothly reduced to alkenes with trialkylammonium formate, usually in high yield.[6] Some advantages of this reduction procedure should be noted. The trialkylammonium formate-palladium reducing system is very simple to use.[20] Clean reduction of vinyl triflates to olefins is observed, no over-reduction being detected. The method is of use in the regioselective synthesis of alkenes and dienes. Ketones, alcohols, ethers, aromatic systems, and presumably a variety of other functional groups are unaffected by the reduction conditions. When the reaction is carried out by using DCOOD, this method allows the regioselective and quantitative introduction of a deuterium atom.

The reaction has been successfully extended to the reduction of aryl triflates to arenes.[21]

Some selected examples of palladium-catalyzed reduction of vinyl and aryl triflates are summarized in the Table.

1. Dipartimento di Studi di Chimica e Tecnologia delle Sostanze Biologicamente Attive, Università degli Studi "La Sapienza", P.le A. Moro 5, 00185 Roma, Italy.

2. Dipartimento di Studi Farmaceutici, Università degli Studi "La Sapienza", P.le A. Moro 5, 00185 Roma, Italy.

3. O'Connor, G. L.; Nace, H. R. *J. Am. Chem. Soc.* **1952**, *74*, 5454.

4. Anderson, A. G.; Stang, P. J. *Org. Synth.* **1981**, *60*, 34.

5. Stang, P. J.; Deuber, T. E. *Org. Synth.* **1974**, *54*, 79.

6. Cacchi, S.; Morera, E.; Ortar G. *Tetrahedron Lett.* **1984**, *25*, 4821.

7. Shapiro, R. H. *Org. React.* **1976**, *23*, 405, and references cited therein.

8. Larson, G. L.; Hernandez, E.; Alonzo, C.; Nieves, I. *Tetrahedron Lett.* **1975**, 4005; Pino, P.; Lorenzi, G. P. *J. Org. Chem.* **1966**, *31*, 329.

9. Caglioti, L.; Cainelli, G.; Maina, G.; Selva, A. *Gazz. Chim. Ital.* **1962**, *92*, 309; *Chem. Abstr.* **1962**, *57*, 12572C.

10. Lewis, J. W.; Lynch, P. P. *Proc. Chem. Soc. London* **1963**, 19; Coulter, J. M.; Lewis, J. W.; Lynch, P. P. *Tetrahedron* **1968**, *24*, 4489.

11. Ireland, R. E.; Pfister, G. *Tetrahedron Lett.* **1969**, 2145; Fetizon, M.; Jurion M.; Anh, N. T. *J. Chem. Soc., Chem. Commun.* **1969**, 112; Ireland, R. E.; Kowalski, C. J.; Tilley, J. W.; Walba, D. M. *J. Org. Chem.* **1975**, *40*, 990; Ireland, R. E.; Muchmore, D. C.; Hengartner, U. *J. Am. Chem. Soc.* **1972**, *94*, 5098; Ireland, R. E.; O'Neil, T. H.; Tolman, G. L. *Org. Synth.* **1983**, *61*, 116.

12. Welch, S. C.; Walters, M. E. *J. Org. Chem.* **1978**, *43*, 2715.

13. Ben-Efraim, D. A.; Sondheimer, F. *Tetrahedron* **1969**, *25*, 2823; Fishman, J.; Torigoe, M.; Guzig, H. *J. Org. Chem.* **1963**, *28*, 1443.

14. Trost, B. M.; Ornstein, P. L. *Tetrahedron Lett.* **1981**, *22*, 3463.

15. (a) Cacchi, S.; Morera, E.; Ortar, G. *Tetrahedron Lett.* **1984**, *25*, 2271; (b) Harnisch, W.; Morera, E.; Ortar, G. *J. Org. Chem.* **1985**, *50*, 1990; (c) Scott, W. J.; Peña, M. R.; Swärd, K.; Stoessel, S. J.; Stille, J. K. *J. Org. Chem.* **1985**, *50*, 2302; (d) Arcadi, A.; Marinelli, F.; Cacchi, S. *J. Organomet. Chem.* **1986**, *312*, C27; (e) Cacchi, S.; Ciattini, P. G.; Morera, E.; Ortar, G. *Tetrahedron Lett.* **1987**, *28*, 3039.

16. Stang, P. J.; Treptow, W. *Synthesis* **1980**, 283; Hassdenteufel, J. R.; Hanack, M. *Tetrahedron Lett.* **1980**, *21*, 503; Stang, P. J.; Fisk, T. E. *Synthesis* **1979**, 438.

17. Cacchi, S.; Morera, E.; Ortar, G. *Tetrahedron Lett.* **1985**, *26*, 1109.

18. Cacchi, S.; Morera, E.; Ortar, G. *Synthesis* **1986**, 320.

19. Stille, J. K.; Tanaka, M. *J. Am. Chem. Soc.* **1987**, *109*, 3785.

20. Weir, J. R.; Patel, B. A.; Heck, R. F. *J. Org. Chem.* **1980**, *45*, 4926; Cacchi, S.; Palmieri, G. *Synthesis* **1984**, 575; Arcadi, A.; Cacchi, S.; Marinelli, F. *Tetrahedron Lett.* **1986**, *27*, 6397; Tsuji, J.; Sugiura, T.; Minami, I. *Synthesis* **1987**, 603.

21. (a) Cacchi, S.; Ciattini, P. G.; Morera, E.; Ortar, G. *Tetrahedron Lett.* **1986**, *27*, 5541; (b) Chen, Q.-Y.; He, Y.-B.; Yang, Z.-Y. *J. Chem. Soc., Chem. Commun.* **1986**, 1452; (c) Peterson, G. A.; Kunng, F.-A.; McCallum, J. S.; Wulff, W. D. *Tetrahedron Lett.* **1987**, *28*, 1381.

TABLE

Palladium-Catalyzed Reduction of Vinyl and Aryl Triflates

Substrate	Catalyst	Product	(% yield)
	Pd(OAc)$_2$(PPh$_3$)$_2$		(81)[6]
	" "		(93)[6]
	" "		(85)[6]
	" "		(95)[6]
	" "		(87)[6]
	Pd(OAc)$_2$/DPPF[b]		(79)[21a]
	" "		(94)[21a]

Substrate	Catalyst	Product	(% yield)
TfO	Pd(OAc)$_2$/DPPF[b]	D	(87)[21a]
N OTf	" "	N	(82)[21a]
TfO	Pd(OAc)$_2$(PPh$_3$)$_2$		(91)[21a]

a) DCOOD was used. b) DPPF refers to 1,1'-bis(diphenylphosphino)ferrocene.

Appendix

Chemical Abstracts Nomenclature (Collective Index Number);

(Registry Number)

Cholesta-3,5-diene (8,9); (747-90-0)

Cholesta-3,5-dienyl trifluoromethanesulfonate: Cholesta-3,5-dien-3-ol,
trifluoromethanesulfonate (11); (95667-40-6)

2,6-Di-tert-butyl-4-methylpyridine: Pyridine, 2,6-bis(1,1-dimethylethyl)-
4-methyl- (9); (38222-83-2)

Trifluoromethanesulfonic anhydride: Methanesulfonic acid, trifluoro-,
anhydride (8,9); (358-23-6)

Cholesten-4-en-3-one (8,9); (601-57-0)

Tributylamine (8); 1-Butanamine, N,N-dibutyl- (9); (102-82-9)

Palladium acetate: Acetic acid, palladium (2+) salt (8,9); (3375-31-3)

Triphenylphosphine: Phosphine, triphenyl- (8,9); (603-35-0)

2-(PHENYLSULFONYL)-1,3-CYCLOHEXADIENE

(Benzene, (1,5-cyclohexadien-1-ylsulfonyl)-)

Submitted by Jan-E. Bäckvall, Seppo K. Juntunen, and Ove S. Andell.[1]

Checked by Willi-Kurt Gries and Larry E. Overman.

1. Procedure

A. trans-3-(Phenylsulfonyl)-4-(chloromercuri)cyclohexene. A 1-L, one-necked, round-bottomed flask equipped with a large magnetic stirring bar is charged with 32.6 g (120 mmol) of mercury(II) chloride (Note 1), 24.6 g (150 mmol) of sodium benzenesulfinate (Note 2), 80 mL of dimethyl sulfoxide and 400 mL of water. The slurry is initially stirred at room temperature for 2 hr and then 10.6 g (132 mmol) of 1,3-cyclohexadiene (Note 3) is added dropwise under vigorous stirring at room temperature over a period of a few minutes. The reaction mixture is stirred for another 2 hr. The reaction flask is cooled with ice and the solid material collected by filtration using a Büchner funnel (Note 4), washed with 400 mL of water, and dried in a desiccator over calcium chloride at reduced pressure (oil pump), to give 53.0 g (97%) of essentially pure trans-3-(phenylsulfonyl)-4-(chloromercuri)cyclohexene (Note 5).

B.   *2-(Phenylsulfonyl)-1,3-cyclohexadiene.*   A 1-L, one-necked round-bottomed flask, equipped with a magnetic stirring bar is charged with 53.0 g (116 mmol) of trans-3-(phenylsulfonyl)-4-(chloromercuri)cyclohexene and 600 mL of diethyl ether (Note 6) at room temperature. The slurry is stirred for 5 min (Note 7) and 175 mL (350 mmol) of a 2 M aqueous solution of sodium hydroxide is added under vigorous stirring (Note 8). The reaction mixture immediately turns black and the vigorous stirring is continued for 30 min (Note 9). The two layers are separated and the aqueous phase is extracted three times with 50-mL portions of diethyl ether. The combined organic layers are filtered through a short column containing 10 g of silica gel and the column is washed with 250 mL of diethyl ether. The ethereal solution is dried over anhydrous magnesium sulfate and filtered, and the solvent is removed at reduced pressure using a rotary evaporator to give 22.5-24.5 g (88-96%) of 2-(phenylsulfonyl)-1,3-cyclohexadiene as a colorless solid (Note 10).

2. Notes

1.   Mercury(II) chloride was purchased from Merck & Company, Inc. and used as delivered. The checkers used material purchased from Mallinckrodt Inc.

2.   Sodium benzenesulfinate (benzenesulfinic acid, sodium salt) was purchased from Aldrich Chemical Company, Inc. and used without further purification.

3.   1,3-Cyclohexadiene was obtained from Fluka Chemical Corporation and distilled before use. The distillation was performed at ambient temperature and reduced pressure (60-70 mm) and the diene was collected in a flask cooled with liquid nitrogen. The checkers used diene purchased from Aldrich Chemical Company, Inc.

4. A funnel with a fine frit must be used.

5. The crude product melts at 119-123°C and is sufficiently pure for use in the next step. Recrystallization from ethyl acetate provides material melting at 128°C (dec). NMR spectral properties are as follows: ^{1}H NMR (250 MHz, CDCl$_3$) δ: 1.92-2.12 (m, 3 H), 2.36-2.51 (m, 1 H), 2.95-3.05 (m, 1 H, H-4), 4.24-4.34 (m, 1 H, H-3), 5.57-5.81 (m, 1 H, =CH), 6.04-6.16 (m, 1 H, =CH), 7.53-7.65 (m, 3 H, ArH), 7.82-7.87 (m, 2 H, ArH); ^{13}C NMR (75 MHz/CDCl$_3$) δ: 26.5, 26.7, 44.1, 66.2, 119.6, 129.1, 129.3, 134.2, 135.4, 136.6.

6. The submitters report that dichloromethane can be used also as the solvent with similar results. This modification was not checked.

7. trans-3-(Phenylsulfonyl)-4-(chloromercuri)cyclohexene is only partly soluble in diethyl ether when the described proportions are used, whereas a clear solution is obtained when dichloromethane is used as solvent.

8. The submitters report that if dichloromethane is used as solvent, 250 mL (500 mmol) of aqueous 2 M sodium hydroxide is added at this point.

9. The submitters report that if dichloromethane is used as solvent, the reaction mixture is stirred for 1.5 hr.

10. The crude product melts at 60-63°C. The spectral properties of 2-(phenylsulfonyl)-1,3-cyclohexadiene are as follows: IR (KBr) cm^{-1}: 3060, 2982, 2923, 2880, 2830, 1585, 1450, 1305, 1150, 1090, 710, 690; ^{1}H NMR (250 MHz/CDCl$_3$) δ: 2.11-2.21 (m, 2 H), 2.35-2.45 (m, 2 H), 5.90-5.97 (m, 1 H, H-4), 6.07 (ddd, 1 H, J = 9.9, 3.6, 1.8, H-3), 6.91-6.95 (m, 1 H, H-1), 7.47-7.62 (m, 3 H, ArH), 7.84-7.88 (m, 2 H, ArH); ^{13}C NMR (75 MHz/CDCl$_3$) δ: 20.7, 22.3, 118.4, 127.7, 129.1, 130.0, 133.2, 134.8, 138.6, 139.8.

3. Discussion

This procedure[2,3] illustrates a highly selective and facile method for introducing a phenylsulfonyl group into the 2-position of 1,3-diene systems by using commercially available starting materials. The method can be applied to cyclic as well as acyclic systems giving 2-(phenylsulfonyl)-1,3-dienes. In an alternative synthesis[4] via condensation of allyl sulfone with aldehyde and subsequent acylation-elimination, the 2-(phenylsulfonyl)-1,3-dienes obtained are limited to acyclic systems.

The procedure described has been applied[2] to 4-methyl-1,3-pentadiene, 1,3-pentadiene and 1,3-butadiene (Table). Caution must be taken in the handling of the sulfonyldiene products from the two latter dienes. They must be handled and stored in solution since they readily undergo Diels-Alder dimerization when concentrated. For the preparation of 2-(phenylsulfonyl)-1,3-pentadiene, final removal of solvent is never effected, giving a 10 to 50 mM solution of product in the preferred solvent. The solution can be stored at -20°C for several days (<5% dimerization), but the product was usually used within a few hours.

Phenylsulfonyl-1,3-dienes are versatile synthetic intermediates. They can participate in cycloaddition reactions and Michael-type additions[3,5] leading to adducts which can be further functionalized.[3] In the latter case the resulting allylic sulfone can be functionalized by electrophiles, nucleophiles, or both (Figure 1).

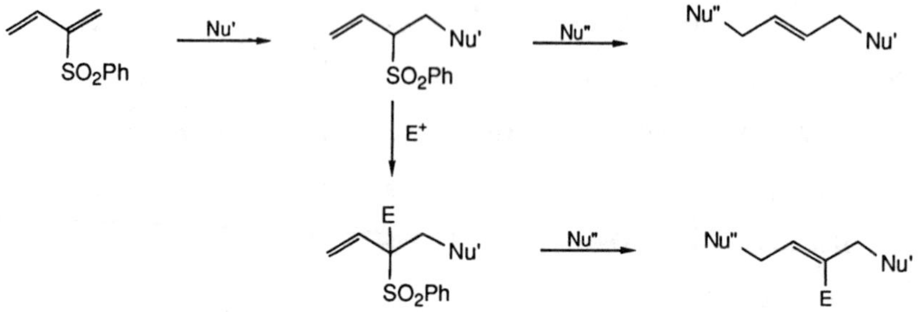

Figure 1

Electron-deficient 1,3-dienes are potentially interesting Diels-Alder dienes. In our study with different kinds of olefins, we observed that 2-(phenylsulfonyl)-1,3-dienes show a duality in their Diels-Alder cycloaddition reactions, giving [4+2] adducts with both electron-deficient and electron-rich olefins.[3] This dual reactivity of the 2-(phenylsulfonyl)-1,3-dienes in [4+2] cycloaddition increases the role they can play in organic synthesis.

1. Department of Organic Chemistry, University of Uppsala, Box 531, S-751 21 Uppsala, Sweden.

2. Andell, O. S.; Bäckvall, J.-E. *Tetrahedron Lett.* **1985**, *26*, 4555.

3. (a) Bäckvall, J.-E.; Juntunen, S. K. *J. Org. Chem.* **1988**, *53*, 2398; (b) Bäckvall, J.-E.; Juntunen, S. K. *J. Am. Chem. Soc.* **1987**, *109*, 6396.

4. (a) Cuvigny, T.; Herve du Penhoat, C.; Julia, M. *Tetrahedron* **1986**, *42*, 5329; (b) Cuvigny, T.; Herve du Penhoat, C.; Julia, M. *Tetrahedron Lett.* **1983**, *24*, 4315.

5. (a) Eisch, J. J.; Galle, J. E.; Hallenbeck, L. E. *J. Org. Chem.* **1982**, *47*, 1608; (b) Overman, L. E.; Petty, C. B.; Ban, T.; Huang, G. T. *J. Am. Chem. Soc.* **1983**, *105*, 6335.

Appendix
Chemical Abstracts Nomenclature (Collective Index Number)
(Registry Number)

2-(Phenylsulfonyl)-1,3-cyclohexadiene: Benzene, (1,5-cyclohexadien-1-ylsulfonyl)- (11); (102860-22-0)

trans-3-(Phenylsulfonyl)-4-(choromercuri)cyclohexene: Mercury, chloro[2-(phenylsulfonyl)-3-cyclohexen-1-yl]-, trans- (11); (102815-53-2)

Sodium benzenesulfinate: Benzenesulfinic acid, sodium salt (8,9); (873-55-2)

1,3-Cyclohexadiene (8,9); (592-57-4)

Table

2-(Phenylsulfonyl) 1,3-Dienes from 1,3-Dienes

Olefin	Sulfonyl Diene	Yield (%)
	SO$_2$Ph	93[a]
	SO$_2$Ph	67[a]
	SO$_2$Ph	93[a]
	SO$_2$Ph	62[b]

[a] From ref. 2 and 3. [b] A modified procedure, compared to that in ref. 2, was used. Acetone was used as solvent in the first step and the reaction time was longer.

TRANSESTERIFICATION OF METHYL ESTERS OF AROMATIC AND α,β-UNSATURATED ACIDS WITH BULKY ALCOHOLS: (-)-MENTHYL CINNAMATE AND (-)-MENTHYL NICOTINATE

(2-Propenoic acid, 3-phenyl-, 5-methyl-2-(1-methylethyl)cyclohexyl ester, [1R-(1α,2β,5α)-)

and

(3-Pyridinecarboxylic acid, 5-methyl-2-(1-methylethyl)cyclohexyl ester, [1R-(1α,2β,5α)]-)

Submitted by Otto Meth-Cohn.[1]

Checked by Gladys Zenchoff, Hubert Maehr, and David Coffen.

1. Procedure

An oven-dried, 500-mL, three-necked, round-bottomed flask is equipped with a magnetic stirring bar, rubber septum inlet, an alcohol thermometer and, through the center neck, a pressure-equalizing dropping funnel bearing a calcium chloride drying tube. The apparatus is flushed with argon (Note 1) and the flask is placed in an ice water bath. To the flask are added (-)-menthol (Note 2) (15.63 g, 100 mmol) and 150 mL of dry tetrahydrofuran (THF) (Note 3). To this stirred solution is added dropwise through the dropping funnel butyllithium in hexane (Note 4) (1.60 M, 55 mL, 88 mmol), transferred by syringe, at such a rate that the temperature does not rise above 20°C (about 10 min). When the addition is complete, the methyl ester [either methyl cinnamate (Note 5) (16.21 g, 100 mmol) or methyl nicotinate (Note 5) (13.71 g, 100 mmol)], dissolved in 30 mL of tetrahydrofuran, is added in one lot to the solution and washed in with an additional 20 mL of tetrahydrofuran. The resulting solution, which slowly becomes cloudy, is stirred for another hour (Notes 6 and 7) and then poured into 200 mL of water in a 1-L separatory funnel; the flask is washed out with 100 mL of diethyl ether. The aqueous layer is separated and the organic phase is washed twice with 200 mL of water. After the organic phase is dried with magnesium sulfate, the solvent is removed on a rotary evaporator and the residue is distilled from a 100-mL flask bearing a Vigreux column (Note 8) under reduced pressure (0.1-0.5 mm). After a forerun of 2-3 g (Note 9) the (1R)-(-)-menthyl ester is collected. (1R)-(-)-Menthyl cinnamate distills at 145-147°C (0.2 mm) and is greater than 99% pure by GLC (Note 10). The yield is 22.6-23.9 g (79-83%). (1R)-(-)-Menthyl nicotinate boils at 141-143°C (0.5 mm) and is greater than 99% pure by GLC (Note 10). The yield is 20.1-21.7 g (77-83%).

156

2. Notes

1. The submitter used a balloon attached by 1" of rubber pressure tubing to the barrel of a plastic disposable syringe bearing a 2"-needle through the septum.

2. (-)-Menthol was obtained from Fluka Chemical Corporation (>99%, puriss. grade) and used directly.

3. Tetrahydrofuran was obtained from BDH Chemicals Ltd. and was distilled from sodium and benzophenone.

4. Butyllithium in hexane was purchased from Lithium Corporation of Europe. The checkers used material supplied by the Aldrich Chemical Company, Inc.

5. Methyl cinnamate (>99%) and methyl nicotinate (>99%) were used as supplied by Fluka Chemical Corporation.

6. The reactions may be monitored by TLC. The submitter used Merck Silica gel pre-coated plates, Silica gel 60 F-254, employing diethyl ether:hexane (1:5) for the cinnamate and (1:1) for the nicotinate transesterifications. After 5 min the reactions are already largely complete. On a small scale, purification by flash chromatography is most effective. R_f times of the reactants are as follows: methyl cinnamate, 0.47; menthyl cinnamate, 0.73; methyl nicotinate, 0.26; menthyl nicotinate, 0.47. Menthol was not visible under ultraviolet light as were the esters, but may be visualized with iodine or phosphomolybdic acid and heat: R_f 0.22 (1:4 ether:hexane).

7. The ice bath may be left in place after all of the reactants are added since the transesterifications are rapid, even below 10°C. If the reaction time is prolonged, only a small difference in product yields results.

8. The submitter used a 120 x 20-mm Vigreux column; the checkers used a 120 x 10-mm column.

9. The forerun contained a mixture of menthol and the methyl ester.

10. Capillary GLC analysis (50 m, OV17, He) gave the following retention times: Menthyl cinnamate 17.5 min (programmed 100-240°C, 5°C/min); menthyl nicotinate 20.2 min (programmed 150-240°C, 5°C/min). The products showed the following properties: Menthyl cinnamate: $[\alpha]_D^{20}$ -57.8° (CHCl$_3$, c 0.20) [lit.[2] $[\alpha]_D^{25}$ -59.5° (CHCl$_3$, c 7.5)]; ^{1}H NMR (CDCl$_3$) δ: 0.6-2.25 (m, 18 H, aliphatic), 4.74 (dt, 1 H, J = 9.5 and 5, O-CH), 6.35 (d, 1 H, J = 17, olefinic), 7.2-7.6 (m, 5 H, aromatic), 7.83 (d, 1 H, J = 17, olefinic). Menthyl nicotinate: $[\alpha]_D^{20}$ -86.8° (CHCl$_3$, c 0.11); ^{1}H NMR (CDCl$_3$) δ: 0.6-2.35 (m, 18 H, aliphatic), 5.01 (dt, 1 H, J = 4.5 and 10.0, O-CH), 7.40 (ddd, 1 H, J = 0.8, 4.8 and 7.8, H-5), 8.32 (dt, 1 H, J = 1.8 and 7.8, H-4), 8.81 (dd, 1 H, J = 1.8 and 4.8, H-6), 9.30 (dd, 1 H, J = 0.8 and 2.1, H-2).

The optical rotations of the products showed a marked dependence on concentration. The submitters found $[\alpha]_D^{20}$ -60.7° (CHCl$_3$, c 0.11) for menthyl cinnamate and $[\alpha]_D^{20}$ -87.9° (CHCl$_3$, c 0.11) for menthyl nicotinate.

3. Discussion

The method described here is based on the general method for such transesterifications.[3] The best alcohol is bulky or tertiary, a feature disfavored by most other methods. Thus tert-butyl alcohol, tert-amyl alcohol, lanosterol, cholesterol, fenchol, and borneol are highly effective. If primary alcohols (e.g., allyl alcohol) are used, it is better to employ 3-5 equiv for an efficient reaction. Alcohols bearing other hetero atoms which form complexes with lithium (e.g., carbohydrate derivatives) prove ineffective in the transesterification.

Methyl esters are always the preferred substrates, conversions being lower with, for example, ethyl esters. Functional groups such as nitro, methoxy, alkenyl and pyridyl are compatible with the reaction conditions. Diesters can only be effective if bis-transesterification is desired, when an excess of the alcohol (e.g., 3-5 equiv) is necessary. Methyl acrylate tends to polymerize under the reaction conditions, but the use of an excess of the ester (3-5 equiv) and lower temperatures (-10°C) allows efficient isolation of the required ester.

Organolithium compounds other than butyllithium can be used with no change in the reaction efficiency; reduction of the molar ratio of organolithium to alcohol merely slows the transesterification. Even when one-sixth of an equivalent is used, efficient but slow transesterification occurs. In no case has it been found necessary to leave reactions longer than 18 hr or to use temperatures higher than ambient. Ether solvents are far more effective than hydrocarbons, in which slower reactions occur.

There are very few known methods for transesterifications using bulky alcohols. Thiol esters undergo ready mercury(II) trifluoroacetate-catalyzed transesterifications with tert-butyl alcohol.[4] Potassium tert-butoxide in the presence of 4-Å molecular sieves converts certain dimethyl malonates into methyl tert-butyl malonates.[5] The majority of published transesterification methods involve the use of primary or occasionally secondary alcohols and a catalyst, and either require a large excess of one reactant or continuous removal of a low boiling component in the equilibrium. Catalysts include acids such as sulfuric[6] or p-toluenesulfonic acid,[7] Lewis acids such as boron tribromide,[8] or bases such as alkoxides.[4,9] Neutral catalysts, in particular titanates,[10] and potassium cyanide[11] have also been used.

1. National Chemical Research Laboratories, CSIR, P. O. Box 395, Pretoria 0001, South Africa. Present address: Sterling Organics Ltd., Newcastle-on-Tyne, England.

2. Sandborn, L. T.; Marvel, C. S. *J. Am. Chem. Soc.* **1926**, *48*, 1409-1413.

3. Meth-Cohn, O. *J. Chem. Soc., Chem. Commun.* **1986**, 695-697.

4. Chan, W. K.; Masamune, S.; Spessard, G. O. *Org. Synth.* **1982**, *61*, 48-55.

5. Wulfman, D. S.; McGiboney, B.; Peace, B. W. *Synthesis*, **1972**, 49.

6. Rehberg, C. E. *Org. Synth., Collect. Vol. III* **1955**, 46-48.

7. Rehberg, C. E. *Org. Synth., Collect. Vol. III* **1955**, 146-148.

8. Yazawa, H.; Tanaka, K.; Kariyone, K. *Tetrahedron Lett.* **1974**, 3995-3996.

9. Braun, G. *Org. Synth., Collect. Vol. II* **1943**, 122-123; Meinwald, J.; Crandall, J.; Hymans, W. E. *Org. Synth., Collect. Vol. V* **1973**, 863-866.

10. Seebach, D.; Hüngerbuhler, E.; Naef, R.; Schnurrenberger, P.; Weidmann, B.; Züger, M. *Synthesis* **1982**, 138-141; Schnurrenberger, P.; Züger, M. F.; Seebach, D. *Helv. Chim. Acta* **1982**, *65*, 1197-1201.

11. Mori, K.; Tominaga, M.; Takigawa, T.; Matsui, M. *Synthesis* **1973**, 790-791.

Appendix

Chemical Abstracts Nomenclature (Collective Index Number);
(Registry Number)

(-)-Menthyl cinnamate: Menthyl cinnamate, (-)- (8); 2-Propenoic acid,
3-phenyl-, 5-methyl-2-(1-methylethyl)cyclohexyl ester, [1R-(1α,2β,5α)]-
(9); (16205-99-5)

(-)-Menthol: Menthol, (-)- (8); Cyclohexanol, 5-methyl-2-(1-methylethyl)-,
[1R-(1α,2β,5α)]- (9); (2216-51-5)

Methyl cinnamate: Cinnamic acid, methyl ester (8); 2-Propenoic acid,
3-phenyl-, methyl ester (9); (103-26-4)

Methyl nicotinate: Nicotinic acid, methyl ester (8); 3-Pyridinecarboxylic
acid, methyl ester (9); (93-60-7)

3-METHYL-2(5H)-FURANONE

(2(5H)-Furanone, 3-methyl-)

A. $\xrightarrow{\text{H}_2\text{O}_2/\text{HCOOH}}$

B. $\xrightarrow{\text{POCl}_3/\text{i-Pr}_2\text{EtN}}$ OPOCl$_2$

C. OPOCl$_2$ $\xrightarrow{\text{Me}_2\text{NH}}$ OPO(NMe$_2$)$_2$

D. OPO(NMe$_2$)$_2$ $\xrightarrow[\substack{\text{2) MeI} \\ \text{3) HCOOH}}]{\text{1) n-BuLi}}$

Submitted by Jan H. Näsman.[1]

Checked by Alan T. Johnson and James D. White.

1. Procedure

Caution. *Hydrogen peroxide attacks the skin and may decompose violently. The first step should be carried out behind a safety screen, and the operator should wear safety glasses and rubber gloves. Air must not be admitted to the hot distillation residue in step 2.*

A. *2(5H)-Furanone.* A 6-L, three-necked, round-bottomed flask equipped with two condensers, a dropping funnel and a 12 x 55 mm magnetic stirring bar is charged with 480 g (5 mol) of furfural (Note 1) and 2.0 L of methylene

162

chloride. The addition of 200 g of sodium sulfate (Note 2) and 150 g of N,N-dimethylethanolamine (Note 3) in one portion each is followed immediately by 460 g of formic acid (Note 4), carefully added in portions over a period of 2 min, after which 100 mL of 30% hydrogen peroxide (Note 5) is added in one portion. The mixture is stirred vigorously. After 5 min the mixture will reflux and another 800 mL of 30% hydrogen peroxide is added dropwise during 9 hr (Note 6) while stirring is continued. When the addition is complete the mixture is vigorously stirred as long as it refluxes and then stirred gently overnight. The organic phase is separated, and the water phase is extracted with the 200 mL of methylene chloride that is used to wash out residues from the reaction flask.

The methylene chloride-phase is washed with two 150-mL portions of saturated sodium disulfite solution (Note 7) and dried over magnesium sulfate and sodium sulfate. After a negative peroxide test (Note 8) the solvent is removed. The crude product (255 g) is fractionated through a 30-cm Vigreux column. The material boiling at 85-95°C (13 mm) is collected to give 210 g of butenolide, which is yellow because of some high boiling residues. Redistillation through the 30-cm Vigreux column and collection of the material boiling at 100-102°C (30 mm) or 95-97°C (19 mm) or 89-91°C (16 mm) or 79-81°C (9 mm), gives colorless butenolide. In this way 170.2 g (41%) of pure butenolide is obtained.

B. *Furyl phosphorodichloridate.* A 1-L flask, protected from moisture by a calcium chloride tube, is charged with 42 g (0.5 mol) of 2(5H)-furanone, 85 g (0.55 mol) of phosphoryl chloride and 100 mL of methylene chloride. A solution of 65 g (0.5 mol) of ethyldiisopropylamine in 60 mL of methylene chloride is added dropwise during 4 hr at ambient temperature (Note 9). The resulting mixture is stirred overnight (12 hr), after which 6.5 g of the amine in 10 mL of methylene chloride is added in one portion and stirring is continued for 20 hr (Note 10). The solvent is removed on a rotary evaporator and 200 mL of dry ether (Note 11) is added cautiously, followed by 100 mL of pentane (in that order), to the dark residue to precipitate the amine hydrochloride. The flask is stoppered and shaken for 1-2 min. The hydrochloride is filtered by suction and washed immediately with 100 mL of dry ether and 200 mL of pentane or petroleum ether (Note 12). The bottle is tightly stoppered and the filtrate is allowed to stand in the refrigerator (+4°C) overnight. The clear brown ethereal phase is decanted from a dark lower phase, and the solvent is evaporated. The residue (~ 100 g) is distilled at the water pump. In order to obtain pure, color-stable, yellow dichloridate it is usually necessary to distill it twice. The first distillation is done rapidly, collecting the material that boils at 73-98°C (9 mm) to give 65-75 g of product, which usually darkens within a few days (Notes 13 and 14). Redistillation (Note 15), collecting the material that boils at 91-93°C (22 mm) or 88-90°C (16 mm) or 73-76°C (9 mm), gives 60-65 g of pure product (Note 16). The yield is 60-65%.

C. *Furyl N,N,N',N'-tetramethyldiamidophosphate.* To 180 mL of dry diethyl ether, chilled to -30°C, is added 56.7 g (4.2 equiv) (1.26 mmol) of dimethylamine (Note 17). This solution is added during 1-2 hr from a double jacketed dropping funnel, protected from moisture by a calcium chloride-tube and connected to a cryostat regulated to -30°C, to a stirred mixture (Note 18) of 60 g (0.30 mol) of the freshly distilled furyl phosphorodichloridate and 250 mL of ether in a two-necked, 1-L flask equipped with a condenser, protected from moisture by a calcium chloride-tube and connected to the cryostat. This flask is chilled in an ice bath during the addition of the first two equivalents of the dimethylamine. After the addition of dimethylamine is complete, stirring is continued for 20 hr while the mixture is warmed on a water bath at 35°C. The hydrochloride which forms is carefully filtered off with suction and washed with two 70-mL portions of dry ether. The combined ether phases are evaporated to give ~65 g (99%) of crude product. Distillation, discarding a yellow forerun and collecting the fraction boiling at 149-152°C (20 mm) or 131-134°C (7 mm) (Note 19), affords 52-58 g (79-88%) of pure material (Note 20).

D.   *3-Methyl-2(5H)-furanone.*   To 10.9 g (50 mmol) of furyl tetramethyldiamidophosphate in 90 mL of tetrahydrofuran (THF) (Note 21), chilled to -75°C, is added 21.9 mL (55 mmol) of a 2.51 M hexane solution of butyllithium (Note 22) at a rate (6-10 min) such that the temperature reaches -60°C but does not exceed this temperature. The resulting mixture is chilled to -75°C for 10 min; then 8.9 g (63 mmol) of methyl iodide in 20 mL of tetrahydrofuran is added with a syringe during 7-8 min (Note 23) so that the temperature does not rise above -55°C. After the addition is complete, the temperature is raised to 0°C and the mixture is concentrated to ca. 40 mL. Water (30 mL) and ethyl acetate (50 mL) are added, the phases are separated, and the dark inorganic phase is extracted with two 50-mL portions of ethyl acetate. The combined, yellow organic phases are washed with brine and dried over magnesium sulfate. The solvent is evaporated to give 10.5 g of crude 2-(3-methylfuryl) tetramethyldiamidophosphate, which need not be purified for the next reaction (Note 24).

To the phosphate in a 250-mL flask on a water bath at 25°C (Note 25) is added 20 mL of 98-100% formic acid (Note 26) and the resulting mixture is stirred until bubbling has ceased (30-40 min). Benzene (50 mL) is added and most of the excess formic acid is removed on an evaporator. To the residue are added 50 mL of ethyl acetate and 30 mL of a sodium chloride-sodium carbonate solution (Note 27). The organic phase is washed twice with the latter solution (i.e., a total of 3 x 30 mL), the combined inorganic phases are extracted once with 50 mL of ethyl acetate, the combined organic phases are dried over magnesium sulfate, the solvent is removed and the product is distilled to give 3.2 g (64%) of 3-methyl-2(5H)-furanone, bp 97-101°C (19 mm).

2. Notes

1. Practical grade furfural from Fluka Chemical Corporation or Aldrich Chemical Company, Inc. was used without any purification. Very dark furfural can be used, but it foams at the beginning of the reaction and leads to lower yields.

2. Sodium sulfate is used to salt-out the water phase; brine is not effective. The yield without the sulfate is 5-10% lower.

3. N,N-Dimethylethanolamine (99% pure) was obtained from EGA CHEMIE or Aldrich Chemical Company, Inc. The role of the compound is to isomerize any 2(3H)-furanone formed.

4. Formic acid (98-100%), obtained from Merck & Company, Inc., was used.

5. "Perhydrol" (30%), obtained from Merck & Company, Inc., gave reproducible results without efforts to determine the activity of the peroxide. An excess is used.

6. The process is a fine balance between oxidation and isomerization of the initially formed 2(3H)-furanone. Longer addition times produce better yields; however, the benefit is of marginal value.

7. Sodium disulfite, $Na_2S_2O_5$, from Merck & Company, Inc. was used. The saturated solution of disulfite should be the lower phase.

8. The mixture is tested for peroxide as follows: Prepare an approximately 1% solution of ferrous ammonium sulfate. Transfer 5 mL to each of two
test tubes and add 0.5 mL of 0.5 M sulfuric acid and 0.5 mL of 0.1 M potassium
thiocyanate solution to each tube.   Add 5 mL of the methylene chloride
solution to one of the test tubes and shake well. The aqueous phase in the
methylene chloride tube should not develop a brown red color when examined
parallel to the blank.

9.   Phosphoryl chloride from Fluka Chemical Corporation or Aldrich
Chemical Company, Inc., and methylene chloride (purum) from Merck & Company,
Inc., were used. Unless the contents of a freshly opened bottle were used,
methylene chloride was distilled from phosphorus pentoxide (20 g/L) before
use.   The amine (Fluka or Aldrich) was distilled from and stored over
potassium hydroxide. The best yields were obtained with once-recovered amine.

10. The reaction

$$\text{(furanone)} \quad + \quad POCl_3/\ i\text{-}Pr_2EtN \quad \rightleftharpoons \quad \text{(furan-OPOCl}_2) \quad + \quad i\text{-}Pr_2EtN \cdot HCl$$

is reversible. In order to obtain pure (97-98%) dichloridate it is essential
to add the 6.5 g of amine after the first equivalent has reacted.

11.   Ordinary diethyl ether is stored over calcium chloride for 36 hr,
filtered, and dried over sodium wire.

12.   The use of more pentane or petroleum ether gave a product of better
stability and purity.

13.   Once-distilled product was usually not color-stable for prolonged
periods.

14. The distillation flask is allowed to cool before air is passed into it. A vigorous polymerization may occur if air is passed into the hot residue, which may be safely discarded after the addition of acetone (an exothermic, but easily controlled reaction).

Spectroscopic data for furyl phosphorodichloridate are as follows: ^{1}H NMR (60 MHz, CDCl$_3$, TMS) δ: 5.85 (m, 1 H, furan-H3), 6.30 (m, 1 H, furan-H4), 7.05 (m, 1 H, furan-H5); ^{13}C NMR (CDCl$_3$, TMS) δ: 92.5 ($^3J_{PC}$ = 7, furan-C3), 111.5 ($^4J_{PC}$ = 3, furan-C4), 137.1 ($^4J_{PC}$ = 3, furan-C5), 147.9 ($^2J_{PC}$ = 12, furan-C2). MS m/e (rel. int.): 202 (16), 200 (26), 119 (4), 117 (6), 83 (100), 55 (31); M$^+$ 201.9160: Calcd 201.9167 for C$_4$H$_3$Cl$_2$O$_3$P; obsv 199.9195: Calcd 199.9197; IR cm^{-1}: 1610 (s), 1300 (s), 980 (s), 890, 870; Anal. Calcd for C$_4$H$_3$Cl$_2$O$_3$P: C, 23.9, H, 1.5; Found: C, 23.8, H, 1.5.

15. Pure pale yellow dichloridate is stable for months without extensive change of color if stored in well-stoppered bottles in the refrigerator.

16. The purity of this product is 97-98%. It contains some butenolide; therefore an excess of dimethylamine is used in the subsequent step.

17. Dry dimethylamine from Fluka Chemical Corporation or MC and B Manufacturing Chemists was used as delivered.

18. A 12 x 55 mm heavy magnetic stirring bar is used for good stirring.

19. The monochloroamidate distills at 123°C (9 mm) and is identified in the ^{1}H NMR by its $^3J_{PH}$ = 13.5. *Distill slowly in the beginning!* The purity of the product is 99+%. Redistill if a dark yellow color develops; however, this color does not preclude successful lithiation.

20. The distilled diamide is a pale yellow oil at room temperature; it freezes in the refrigerator (+4°C) if seeded within some hours. The first spontaneous crystallization took several weeks. It can also be obtained as snow white crystals from diisopropyl ether/hexane, mp 15-16°C.

Spectroscopic data for furyl tetramethyldiamidophosphate are as follows: ^{1}H NMR (400 MHz, CDCl$_3$, TMS) δ: 2.71 (d, 12 H, [3]J$_{PH}$ = 10, two N(CH$_3$)$_2$), 5.62 (m, 1 H, furan-H3), 6.28 (m, 1 H, furan-H4), 6.95 (m, 1 H, furan-H5); ^{13}C NMR (15.03 MHz, CDCl$_3$, TMS) δ: 151.9 (d, [2]J$_{PC}$ = 6, furan-C2), 134.5 (s, furan-C5), 111.3 (s, furan-C4), 88.8 (d, [3]J$_{PC}$ = 4, furan-C3), 36.6 (d, [2]J$_{PC}$ = 4, N(CH$_3$)$_2$).   Note that multiplicities s and d refer to C-P coupling. MS m/e (rel. int.): 218 (6), 136 (6), 135 (100), 127 (2), 111 (2), 92 (7), 90 (2), 83 (4), 69 (3).   M$^+$ at 218.0822:   Calcd 218.0820 for C$_8$H$_{15}$N$_2$O$_3$P; IR cm^{-1}: 2900, 2800, 1610, 1300, 990, 960.

21. Tetrahydrofuran (Merck) was distilled from sodium-benzophenone ketyl prior to use.

22. Butyllithium was obtained from Aldrich Chemical Company, Inc., and methyl iodide from Merck & Company, Inc. Butyllithium was titrated with phenanthroline as indicator prior to use according to the method of Watson and Eastham.[11] Fresh alkoxide-free butyllithium should be used to ensure pure product.

23. Methyl iodide should be added carefully in the beginning when the reaction mixture is mostly solid.

24. The phosphate can be crystallized from diisopropyl ether/hexane at -20°C in 80-85% yield; mp 42-44°C.

Spectroscopic data for 2-(3-methylfuryl) tetramethyldiamidophosphate are as follows: ^{1}H NMR (400 MHz, CDCl$_3$, TMS) δ: 1.95 (dxt, 3 H, J = 0.4 and 2.2 CH$_3$), 2.73 (d, 12 H, [3]J$_{PH}$ = 10.2, two N(CH$_3$)$_2$), 6.16 (dxdxq, 1 H, J = 0.4 and 2.2, furan-H4), 6.91 (dxdxq, 1 H, J = 0.4 and 2.2, furan-H5); ^{13}C NMR δ: 8.4 (sxq, CH$_3$), 36.6 (dxq, [2]J$_{PC}$ = 4, N(CH$_3$)$_2$), 98.7 (dxs, [3]J$_{PC}$ = 5, furan-C3), 113.8 (dxd, [4]J$_{PC}$ = 2, furan-C4), 133.9 (dxd, [4]J$_{PC}$ = 2, furan-C5), 147.8 (dxs, [3]J$_{PC}$ = 8, furan-C2); multiplicities underlined in the ^{13}C-spectrum

refer to C-P coupling, the other to C-H coupling; MS m/e (rel. int.): 232
(7), 135 (100), 97 (3), 92 (5), M$^+$ at 232.0980: Calcd 232.0977 for
$C_9H_{17}N_2O_3P$; Calcd for $C_9H_{17}N_2O_3P$: C, 46.55, H, 7.33, N, 12.07; Found: C,
46.5, H, 7.6, N, 12.0.

25. The water bath can be removed after 5 min. The reaction is vigorous
in the beginning and chilling is necessary to avoid formation of dimethylformamide (DMF), which is formed at elevated temperatures.

26. Formic acid (98-100%), obtained from Merck & Company, Inc., was
used.

27. *Warning*: CO_2 evolution. The sodium chloride-sodium carbonate
solution was prepared from 185 g of sodium chloride and 110 g sodium carbonate
dissolved in water to give a total volume of 1 L.

3. Discussion

The preparation of 2(5H)-furanone is the scaled-up and slightly modified
procedure[2] based on the report of Badovskaya that furfural is oxidized with
performic acid to give a mixture of furanones.[3] The preparation here is
improved by the use of N,N-dimethylaminoethanol as a catalyst for the
isomerization of 2(3H)-furanone to 2(5H)-furanone. The complex between this
amino alcohol and formic acid does not enter the organic phase during workup
and the product is thus easily isolated simply by evaporation of the solvent.

The hitherto preferred method for preparation of the butenolide is that
of Price and Judge,[4] which can be modified (by extraction of the bromolactone
with methylene chloride and elimination of hydrogen bromide with triethylamine
or preferably with diisopropylethylamine in toluene at 70°C) to give routinely
60% or greater overall yield on a 6-mol scale. However, the large amount of

hydrogen bromide evolved is sometimes a nuisance, especially to inexperienced workers. The method reported here is fast, independent of scale (0.1-6 mol tried), the starting materials are cheap, and the product is easily isolated.

Substituted furfurals do not react at a synthetically-useful rate when formic acid/hydrogen peroxide is used. This suggests that the reaction takes place in the water phase and that substituted furfurals enter this phase only with difficulty.

The preparation of furyl phosphorodichloridate is based upon a method to prepare 2-chlorofuran (16% yield, Hormi, Näsman unpublished). Later the preparation was extended to a general method to prepare furyl esters from carboxylic acid chlorides lacking α-hydrogens and alkyl furyl carbonates from primary (other than methyl) and secondary alkyl chloroformates.[5] Phosphoryl chloride was the only acid chloride except carbon analogues found to give a furyl ester by the amine-catalyzed reaction.

Regioselective β-metallation of π-excessive five ring heterocycles is not a novel reaction.[6] Oxazoline[7] and pyridine[8] as well as carboxylate-[9] and carboxamide[10]-substituted heterocycles have been lithiated. From the point of synthetic utility thiophenes have been shown to be useful substrates after careful optimization of reaction conditions; furans have been of less utility.

The generation of 2-(3-lithio)furyl tetramethyldiamidophosphate ($\geq$ 95%) in tetrahydrofuran with a slight excess of butyllithium is a reliable procedure. The reagent usually forms a precipitate (active) when stored for prolonged times (3-12 hr) at -80°C and less than 35 mL of tetrahydrofuran/10 mmol of reagent is used.[6] The reagent darkens when warmed to -30°C. The reagent has been used on 100-mmol scales with no difficulties in methylation with methyl iodide. Methyl iodide is a very good electrophile for the reagent, whereas ethyl iodide does not react. Other good substrates for this

171

unmodified reagent are ketones, aldehydes, and chloromethyl ethers. Alkylation is difficult unless strongly activated substrates are used. For example, benzyl chloride is unreactive, benzyl bromide reacts but not completely, and the corresponding iodide gives a complex mixture.

The reagent must be added to the electrophile when the leaving group is an alkoxide. For example, quenching with MeOD on larger scales yields products labelled also in the 5-position, whereas reverse addition with good stirring does not.

The furan products can be purified by flash chromatography[12] and should be used at once. A mixture of ethyl acetate and methylene chloride is a good solvent system for flash chromatography. Small residues of silica tend to partly decompose these furans within two weeks. The products are hygroscopic. Diisopropyl ether-diethyl ether-hexane is a useful solvent system for recrystallization of solid furans.

The formic acid reaction to convert furans to butenolides seems to be general, although heating may be necessary for acceptor-substituted furans; dimethylformamide (DMF) is then a byproduct.

In conclusion, furyl N,N,N',N'-tetramethylamidophosphate is the precursor to the d^2-synthon 1 for butenolide, which is difficult to generate by direct or other indirect methods;[13] however, see reference 14 for a metal-halogen exchange reaction of 3- or 4-bromo-2-methoxy- or 2-trimethylsiloxyfurans.

1

1. Institutionen för Organisk Kemi, Åbo Akademi, Akademig. 1, 20500 Åbo 50, Finland. I gratefully acknowledge a fellowship from the Academy of Finland.

2. Näsman, J. H.; Hormi, O. E. O. Fin. Pat. appl. 85/2444 20, June 1985; Näsman, J.-A. H.; Pensar, K. G. *Synthesis* **1985**, 786.

3. Badovskaya, L. A. *Khim. Geterotsikl. Soedin* **1978**, (10), 1314; *Chem. Abstr.* **1979**, *90*, 54751x.

4. Price, C. C.; Judge, J. M. *Org. Synth.* **1965**, *45*, 22.

5. Hormi, O. E. O.; Näsman, J. H. *Synth. Commun.* **1986**, *16*, 69.

6. Näsman, J. H.; Kopola, N.; Pensar, G. *Tetrahedron Lett.* **1986**, *27*, 1391.

7. (a) DellaVecchia, L.; Vlattas, I. *J. Org. Chem.* **1977**, *42*, 2649; (b) Chadwick, D. J.; McKnight, M. V.; Ngochindo, R. *J. Chem. Soc., Perkin Trans. I* **1982**, 1343; (c) Ribereau, P.; Queguiner, G. *Tetrahedron* **1984**, *40*, 2107; (d) Carpenter, A. J.; Chadwick, D. J. *J. Chem. Soc., Perkin Trans I* **1985**, 173.

8. Ribéreau, P.; Quéguiner, G. *Tetrahedron* **1983**, *39*, 3593.

9. Carpenter, A. J.; Chadwick, D. J. *Tetrahedron Lett.* **1985**, *26*, 1777.

10. Doadt, E. G.; Snieckus, V. *Tetrahedron Lett.* **1985**, *26*, 1149.

11. Watson, S. C.; Eastham, J. F. *J. Organometal. Chem.* **1967**, *9*, 165-168.

12. Still, W. C.; Kahn, M.; Mitra, A. *J. Org. Chem.* **1978**, *43*, 2923.

13. Posner, G. H.; Kogan, T. P.; Haines, S. R.; Frye, L. L. *Tetrahedron Lett.* **1984**, *25*, 2627.

14. Wiesner, K. *Int. Conf. Chem. Biotechnol. Biol. Act. Nat. Prod., [Proc]*, *1st* **1981**, *1*, 7-27; *Chem. Abstr.* **1982**, *97*, 182734a.

Appendix

Chemical Abstracts Nomenclature (Collective Index Number); (Registry Number)

3-Methyl-2(5H)-furanone: 2-(5H)-Furanone, 3-methyl- (8,9); (22122-36-7)

2(5H)-Furanone (8,9); (497-23-4)

Furfural: 2-Furaldehyde (8); 2-Furancarboxaldehyde (9); (98-01-1)

N,N-Dimethylethanolamine: Ethanol, 2-(dimethylamino)- (8,9); (108-01-0)

Furyl phosphorodichloridate: Phosphorodichloridic acid, 2-furanyl ester (12); (105262-70-2)

Phosphoryl chloride (8,9); (10025-87-3)

Ethyldiisopropylamine: Triethylamine, 1,1'-dimethyl- (8); 2-Propanamine, N-ethyl-N-(1-methylethyl)- (9); (7087-68-5)

Furyl N,N,N',N'-tetramethyldiamidophosphate: Phosphorodiamidic acid, tetramethyl-, 2-furanyl ester (12); (105262-58-6)

2-(3-Methylfuryl) tetramethylamidophosphate: Phosphorodiamidic acid, tetramethyl-, 3-methyl-2-furanyl ester (12) (105262-59-7)

INTRAMOLECULAR CYCLIZATION OF cis,cis-1,5-CYCLOOCTADIENE USING

HYPERVALENT IODINE: BICYCLO[3.3.0]OCTANE-2,6-DIONE

(1,4-Pentalenedione, hexahydro-)

Submitted by Robert M. Moriarty,[1] Michael P. Duncan,[1]

Radhe K. Vaid,[1] and Om Prakash.[2]

Checked by Deng Bing and Ekkehard Winterfeldt.

1. Procedure

A.   *2,6-Diacetoxybicyclo[3.3.0]octane*, 2, (Notes 1 and 2).   An oven-dried, 1-L, round-bottomed flask, equipped with a magnetic stirring bar, reflux condenser, and drying tube (Drierite), is charged with iodosobenzene diacetate (IBD) (100 g, 0.31 mol) and 300 mL of glacial acetic acid. To this stirred mixture, 25 g (0.23 mol) of cis,cis-1,5-cyclooctadiene (COD) is added. The resulting mixture is then heated to reflux for 16 hr (Note 3) at which time the colorless solution has become brown-orange. At the end of this time the acetic acid is evaporated using a rotary evaporator (15 mm). Reduced-pressure distillation (74-84°C/0.060 mm) yields 29.1-30.5 g (56-58%) of 2, 2,6-diacetoxybicyclo[3.3.0]octane, as a pale yellow liquid (lit.[3] bp 84-88°C at 0.2 mm) (Note 4).

B. *Bicyclo[3.3.0]octane-2,6-diol*, 3, (Note 5). An ice-cooled aqueous 10% solution of sodium hydroxide (100 mL) is placed in a 250-mL, round-bottomed flask equipped with a magnetic stirring bar and stopper. To this ice-cooled solution 27.8 g of diacetate 2 (0.123 mol) is added dropwise over a few minutes. The cooled solution is slowly allowed to warm to room temperature (1 hr) and stirring is continued for 15 hr, at which time the colorless solution has become yellow-orange (Note 6).

The reaction mixture is then extracted continuously with ether for 3 days. After extraction the ether is removed by rotary evaporation. The crude viscous liquid which results after evaporation (Note 7) is distilled (Note 8) under reduced pressure (106-111°C/0.06 mm) (lit.[3] bp 90-96°C at 0.3 mm) to yield 14.5-16.2 g (83-93%) of 3, pure bicyclo[3.3.0]octane-2,6-diol, as a yellow viscous liquid (Note 9).

C. *Bicyclo[3.3.0]octane-2,6-dione*, 4, (Note 10). Diol 3, 12.6 g (0.089 mol), is placed in a 250-mL, three-necked, round-bottomed flask equipped with a mechanical stirrer and a reflux condenser. Acetone (125 mL) is added and the mixture is cooled to 0°C. A 2.7-M solution of Jones reagent (Note 11) (70 mL) is slowly added dropwise over 10 min at 0°C. The solution is allowed to warm slowly to room temperature (1 hr) and stirring is continued for an additional 15 hr.

After 15 hr the acetone is removed on a rotary evaporator and water (125 mL) is added. The dark green aqueous mixture is extracted continuously with ether for 3 days. The ether is removed by rotary evaporation which results in a yellow oil. The oil is then distilled under reduced pressure (74-79°C/0.06 mm) to yield analytically pure bicyclo[3.3.0]octane-2,6-dione, 4, (6.4-7.1 g, 52-58%) as a white crystalline solid mp 45-46°C; lit.[4] mp 45.1-46.3°C (Notes 12 and 13).

2. Notes

1. cis,cis-1,5-Cyclooctadiene (COD) and iodosobenzene diacetate (IBD) are purchased from Aldrich Chemical Company, Inc.

2. The diacetate (2) is a mixture of three difficultly separable stereoisomers [the di-exo-diacetate (2a), di-endo-diacetate (2b), and the exo-endo-diacetate (2c)]. The major isomer is the di-exo-diacetate (2a) based on ^{13}C-NMR of the known di-exo-diol, see (Note 9).

OAc	OAc	OAc
OAc	OAc	OAc
2a	2b	2c

3. This solution of iodosobenzene, acetic acid, and cis,cis-1,5-cyclooctadiene should continue to be stirred and should not be allowed to react for more than 20 hr (at refluxing temperature) to prevent decomposition of the product diacetate.

4. The ^{1}H NMR spectrum (CDCl$_3$) is as follows δ: 1.60 (m, 8 H, CH$_2$), 1.97 (s, 6 H, OAc), 2.55 (br, s, 2 H, CH), 4.90 (br, s, 2 H, CHOAc). The IR spectrum (neat) shows a carbonyl peak at 1738 cm^{-1}.

5. This procedure for the preparation of the diol is an adapted version of that by Cantrell and Strasser.[3] It is a superior procedure to that of Crandall and Mayer.[5]

6. The checkers monitored the reaction by TLC using ethyl acetate as the developing solvent.

7. This viscous liquid (3) is easily transferred to a distilling flask by using acetone.

8. The use of a heat gun aids the distillation because the product is extremely viscous.

9. The ^{1}H NMR spectrum (CDCl$_3$) is as follows δ: 1.70 (m, 8 H, C_H$_2$), 2.61 (m, 2 H, C_H), 3.05 (s, 2 H, O_H), 3.90 (m, 2 H, C_HOH). The IR spectrum shows a broad peak at 3500 cm^{-1}. The major peaks in the ^{13}C NMR spectrum (CDCl$_3$) are δ: 27.41 (C-4), 33.81 (C-3), 50.64 (C-1), 79.54 (C-2). The ^{13}C NMR indicates that the major stereoisomer is 3a, the exo, exo-2,6-dihydroxy-cis-bicyclo[3.3.0]octane [lit.[6] ^{13}C NMR δ: 27.8 (C-4), 34.2 (C-3), 51.0 (C-1), 79.9 (C-2)].

3a

10. Other oxidation procedures were used, e.g., pyridinium chlorochromate (Corey's reagent),[7] and dipyridine Cr(VI) oxide (Collins' reagent),[8] but did not produce yields comparable to the Jones method.

11. Jones reagent was prepared by the method in Fieser and Fieser:[9] Dissolve 13.36 g of chromium trioxide in 11.5 mL of concd sulfuric acid, and carefully dilute this cooled solution (0°C) with water to 50 mL.

12. The ^{1}H NMR spectrum (CDCl$_3$) is as follows δ: 2.23 (m, 8 H, CH$_2$), 3.00 (m, 2 H, CH). The IR spectrum (Nujol) shows a carbonyl peak at 1745 cm^{-1}.

13. GLC analysis shows that the product is contaminated by small amounts of diol. If desired, purer material could be obtained by sublimation at 35-40°C/0.01 mm onto a cold finger kept at 0°C.[4]

3. Discussion

The preparation of bicyclo[3.3.0]octane-2,6-dione has been accomplished by intermolecular reactions,[4,10] intramolecular reactions,[3,11] and degradation reactions.[5,12]

Bicyclo[3.3.0]octane-2,6-dione has been known since 1934,[10] but extant procedures for large-scale multi-gram synthesis of this versatile intermediate are cumbersome, except for the recently published results of Hagedorn and Farnum.[4] Whitesell and Matthews[6] have shown that bicyclo[3.3.0]octanes are valuable intermediates for the total synthesis of natural products.

We now report a simple, three-step synthesis of the dione, which uses simple procedures and inexpensive starting materials, to procure multigram amounts of bicyclo[3.3.0]octane-2,6-dione in reasonable yields.

1. Department of Chemistry, University of Illinois at Chicago, Chicago, IL 60607.

2. Department of Chemistry, Kurukshetra University, Kurukshetra (Haryana), India 132119.

3. Cantrell, T. S.; Strasser, B. L. *J. Org. Chem.* **1971**, *36*, 670.

4. Hagedorn, A. A., III; Farnum, D. G. *J. Org. Chem.* **1977**, *42*, 3765.

5. Crandall, J. K.; Mayer, C. F.; *J. Am. Chem. Soc.* **1967**, *89*, 4374.

6. Whitesell, J. K.; Matthews, R. S. *J. Org. Chem.* **1977**, *42*, 3878.

7. Corey, E. J.; Suggs, J. W. *Tetrahedron Lett.* **1975**, 2647.

8. Collins, J. C.; Hess, W. W. *Org. Synth.* **1972**, *52*, 5.

9. Fieser, L. F.; Fieser, M. "Reagents for Organic Synthesis," Vol. 1; Wiley: New York, 1967; pp. 142-144.

10. Ruzicka, L.; Borges de Almeida, A.; Brack, A. *Helv. Chim. Acta* **1934**, *17*, 183.

11. Julia, M.; Colomer, E. *An. Quim.* **1971**, *67*, 199; *Chem. Abstr.* **1971**, *75*, 19766z.

12. McCabe, P. H.; Nelson, C. R. *Tetrahedron Lett.* **1978**, 2819.

Appendix

Chemical Abstracts Nomenclature (Collective Index Number)

(Registry Number)

Bicyclo[3.3.0]octane-2,6-dione: 1,4-Pentalenedione, hexahydro- (8,9);
(17572-87-1)

cis,cis-1,5-Cyclooctadiene: 1,5-Cyclooctadiene, (Z,Z)- (8,9);
(1552-12-1)

2,6-Diacetoxybicyclo[3.3.0]octane: 1,4-Pentalenediol, octahydro-, diacetate
(8,9); (17572-85-9)

Iodosobenzene diacetate: Benzene, (diacetoxyiodo)- (8);
Iodine, bis(acetato-O)phenyl- (9); (3240-34-4)

Bicyclo[3.3.0]octane-2,6-diol: 1,4-Pentalenediol, octahydro- (8);
1,4-Pentalenediol, octahydro-, (1α,3aα,4α,6aα)- (10); (17572-86-0)

(Z)-4-(TRIMETHYLSILYL)-3-BUTEN-1-OL

(3-Buten-1-ol, 4-(trimethylsilyl)-, (Z)-)

A.

$$Me_3Si \equiv \quad OH \xrightarrow[\text{2. ClSiMe}_3]{\text{1. EtMgBr}} Me_3Si \equiv \quad OH$$

B.

$$Me_3Si \equiv \quad OH \xrightarrow[\text{Pd/BaSO}_4\text{/quinoline}]{H_2} \quad Me_3Si \quad OH$$

Submitted by Larry E. Overman, Mark J. Brown, and Stephen F. McCann.[1]

Checked by Ronald C. Newbold and Andrew S. Kende.

1. Procedure

A. Preparation of 4-(trimethylsilyl)-3-butyn-1-ol. A flame-dried, three-necked, 2-L, round-bottomed flask is fitted with a 1-L pressure equalizing addition funnel, a mechanical stirrer, and a nitrogen inlet. The flask is flushed with dry nitrogen and charged with 3-butyn-1-ol (Note 1) (freshly distilled, 31.4 g, 0.448 mol) and 900 mL of anhydrous tetrahydrofuran (Note 2). The stirred solution is cooled to 0°C under nitrogen and to it is added over 1 hr a solution of ethylmagnesium bromide in tetrahydrofuran (493 mL of 2.0 M soln, 0.986 mol) (Note 1). The resulting heterogeneous mixture is rapidly stirred at 0°C for 1 hr, allowed to warm to room temperature for 1 hr, and then recooled to 0°C. To this mixture is slowly added over 30 min with rapid stirring freshly distilled (Note 3) chlorotrimethylsilane (125 mL, 0.986 mol). The mixture is stirred for 1 hr at 0°C and allowed to warm to room temperature over 1-2 hr. The entire reaction mixture is poured slowly with

182

rapid stirring into a 4-L Erlenmeyer flask which contains 1 L of ice-cold 3 M hydrochloric acid, and stirred at 25°C for an additional 2 hr. The organic phase is separated and the aqueous phase is extracted with three 200-mL portions of ether.

The combined organic phases are washed with two 200-mL portions of water, four 200-mL portions of saturated sodium bicarbonate solution, and two 200-mL portions of saturated sodium chloride. The organic phase is dried over anhydrous magnesium sulfate, filtered, and concentrated under reduced pressure at room temperature using a rotary evaporator. The crude product is distilled through a short-path distillation apparatus under reduced pressure to give 45.2 g (0.318 mol, 71% yield) of 4-(trimethylsilyl)-3-butyn-1-ol, bp 78-79°C (10 mm), as a colorless liquid (Notes 4 and 5).

B. *Preparation of (Z)-4-(trimethylsilyl)-3-buten-1-ol.* A dry, 250-mL, round-bottomed flask with a stirring bar is charged with 8.84 g (0.062 mol) of 4-(trimethylsilyl)-3-butyn-1-ol, 0.4 g of 5% palladium on barium sulfate (Note 6), 0.45 g of synthetic quinoline (Note 7) and 78 mL of methanol. The flask is placed on a hydrogenation apparatus equipped with a gas burette, and the stirred mixture is thoroughly purged with nitrogen. The nitrogen is then replaced by hydrogen and the reaction mixture is stirred at atmospheric pressure and room temperature until 1.46 L (0.065 mol) of hydrogen is consumed. The flask is flushed with nitrogen and the solution is filtered through a thick pad of Celite. The filtrate is concentrated on a rotary evaporator at room temperature to afford 10-15 mL of an oil, which is diluted with 150 mL of ether. The ether solution is thoroughly washed once with 200 mL of ice-cold 0.2 M sulfuric acid, then once with 20 mL of 5% sodium bicarbonate solution. The ether layer is dried over anhydrous magnesium sulfate, filtered, and concentrated to yield 8.4 g (0.058 mol) of the crude buten-1-ol (Note 8). Short path distillation under reduced pressure gives 7.60 g (0.0527 mol, 85% yield) of (Z)-4-(trimethylsilyl)-3-buten-1-ol, bp 95-100°C (25 mm) as a colorless liquid (Notes 9 and 10).

2. Notes

1. The reagents, 3-butyn-1-ol and 2.0 M ethylmagnesium bromide in tetrahydrofuran, were purchased from Aldrich Chemical Company, Inc. The ethylmagnesium bromide concentration can be easily checked by titration with menthol using 1,10-phenanthroline as indicator.[2]

2. Tetrahydrofuran was distilled from sodium and benzophenone under a nitrogen atmosphere.

3. Chlorotrimethylsilane was purchased from Aldrich Chemical Company, Inc., and was distilled from calcium hydride under an atmosphere of nitrogen immediately prior to use.

4. The product, 4-(trimethylsilyl)-3-butyn-2-ol, shows the following proton NMR spectrum at 300 MHz in $CDCl_3$ δ: 0.03 (s, 9 H, $SiCH_3$), 1.8 (broad s, 1 H, OH), 2.47 (t, 2 H, CH_2), 3.67 (m, 2 H, CH_2OH); and infrared spectrum (neat) cm^{-1}: 3350 (very broad), 2178, 1250, 1031, 894, 842, 760.

5. Similar yields can be obtained in this silylation by using the chloromagnesium salt (from butylmagnesium chloride) as described in *Org. Synth.* **1987**, *65*, 61.

6. The 5% palladium on barium sulfate was purchased from Engelhard Industries, Newark, NJ.

7. Synthetic quinoline was purchased from Aldrich Chemical Company, Inc. and was distilled prior to use.

8. The proton NMR spectrum of the crude buten-1-ol was essentially identical to that of the distilled product, except for traces of solvent. This crude silylbuten-1-ol was of sufficient purity for the tetrahydropyridine synthesis described in the next procedure.

9. Distilled product showed a proton NMR at 250 MHz in $CDCl_3$ as follows δ: 0.14 (s, 9 H, $SiCH_3$), 1.61 (broad s, 1 H, OH), 2.37-2.46 (m, 2 H, CH_2CH_2OH), 3.68 (broadened t, 2 H, J = 6.5, CH_2OH), 5.66-5.71 (dt, 1 H, J = 14.1, J = 1.2, $Me_3SiCH=CHR$), 6.29 (overlapping dt, 1 H, J = 14.1, J = 7.1, $R_3SiCH=CHR$). Gas chromatographic analysis using a 25-m 5% methylphenyl-silicone column showed that this sample was a 92:8 mixture of Z and E isomers and contained <2% of other impurities.

10. The submitters report that Z-4-(trimethylsilyl)-3-buten-1-ol of >98% isomeric purity can be obtained in ca. 60% overall yield by a more lengthy sequence involving hydroalumination-protonolysis[3] of the tetrahydropyranyl (THP) ether of 4-(trimethylsilyl)-3-butyn-1-ol[4] followed by cleavage[5] of the THP ether with pyridinium p-toluenesulfonate in methanol. This sequence is less convenient for the tetrahydropyridine synthesis described in the next procedure, since the isomeric purity of the vinylsilane is not important for the cyclization reaction.[6]

3. Discussion

The direct silylation of 3-butyn-1-ol follows the Danheiser modification[7] of the Westmuze-Vermeer[8] method. The subsequent semihydrogenation is a modification[9] of the Lindlar procedure and yields the Z-alkene isomer in >90% isomeric purity.

The following *Organic Syntheses* procedure[10] illustrates one[6] of the uses of the 4-carbon organosilane intermediates described in this preparation.

1. Department of Chemistry, University of California, Irvine, CA 92717.

2. Watson, S. C.; Eastham, J. F. *J. Organomet. Chem.* **1967**, *9*, 165; Bergbreiter, D. E.; Pendergrass, E. *J. Org. Chem.* **1981**, *46*, 219.

3. Hammoud, A.; Descoins, C. *Bull. Soc. Chim. Fr.* **1978**, 299.

4. Zweifel, G.; Lewis, W. *J. Org. Chem.* **1978**, *43*, 2739.

5. Miyashita, M.; Yoshikoshi, A.; Grieco, P. A. *J. Org. Chem.* **1977**, *42*, 3772.

6. Flann, C.; Malone, T. C.; Overman, L. E. *J. Am. Chem. Soc.* **1987**, *109*, 6097.

7. Danheiser, R. L.; Carini, D. J.; Fink, D. M.; Basak, A. *Tetrahedron* **1983**, *39*, 935.

8. Westmuze, H.; Vermeer, P. *Synthesis* **1979**, 390.

9. Cram, D. J.; Allinger, N. L. *J. Am. Chem. Soc.* **1956**, *78*, 2518.

10. Overman, L. E.; Flann, C. J.; Malone, T. C. *Org. Synth.* **1988**, *68*.

Appendix

Chemical Abstracts Nomenclature (Collective Index Number);

(Registry Number)

(Z)-4-(Trimethylsilyl)-3-buten-1-ol: 3-Buten-1-ol, 4-(trimethylsilyl)-,
(Z)- (11); (87682-77-7)

4-(Trimethylsilyl)-3-butyn-1-ol: 3-Butyn-1-ol, 4-(trimethylsilyl)- (9);
(2117-12-6)

3-Butyn-1-ol (8,9); (927-74-2)

Ethylmagnesium bromide: Magnesium, bromoethyl- (9); (925-90-6)

Chlorotrimethylsilane: Silane, chlorotrimethyl- (8,9); (75-77-4)

Quinoline (8,9); (91-22-5)

REGIOSELECTIVE SYNTHESIS OF TETRAHYDROPYRIDINES:
1-(4-METHOXYPHENYL)-1,2,5,6-TETRAHYDROPYRIDINE

A. Me_3Si—CH=CH—CH$_2$CH$_2$—OH + p-CH$_3$C$_6$H$_4$SO$_2$Cl $\longrightarrow$ Me_3Si—CH=CH—CH$_2$CH$_2$—OSO$_2$C$_6$H$_4$CH$_3$-p

B. Me_3Si—CH=CH—CH$_2$CH$_2$—OSO$_2$C$_6$H$_4$CH$_3$-p + p-CH$_3$OC$_6$H$_4$NH$_2$ $\longrightarrow$ Me_3Si—CH=CH—CH$_2$CH$_2$—NH—C$_6$H$_4$—OMe

C. Me_3Si—CH=CH—CH$_2$CH$_2$—NH—C$_6$H$_4$—OMe $\xrightarrow[\text{p-CH}_3\text{C}_6\text{H}_4\text{SO}_3\text{H}]{(\text{CH}_2\text{O})_n}$ CH$_3$O—C$_6$H$_4$—N(tetrahydropyridine ring)

Submitted by Larry E. Overman, Chris J. Flann, and Thomas C. Malone.[1]
Checked by Ronald C. Newbold and Andrew S. Kende.

1. Procedure

A. Preparation of (Z)-4-(trimethylsilyl)-3-butenyl 4-methylbenzenesulfonate. An oven-dried, 1-L, round-bottomed flask is equipped with a magnetic stirring bar and purged with dry argon or nitrogen. The flask is charged with 17.3 g (0.120 mol) of (Z)-4-(trimethylsilyl)-3-buten-1-ol (Note 1) and 290 mL of dry pyridine (Note 2). The reaction mixture is cooled to 0°C in an ice-water bath and 25.2 g (0.132 mol) of p-toluenesulfonyl chloride (Note 3) is added to the solution. When the p-toluenesulfonyl chloride is completely dissolved, the flask containing the reaction mixture is sealed and placed in a refrigerator at -20°C for 24 hr (Note 4). The reaction mixture is then poured into a rapidly stirring mixture of 200 g of ice and 200 mL of water contained in a 2-L Erlenmeyer flask. The resulting mixture is transferred to a 2-L separatory funnel and extracted with five 200-mL portions of ether. The combined organic phases are washed with five 200-mL portions of ice-cold aqueous 6 N hydrochloric acid (Note 5) and 200 mL of water. The organic phase is dried over anhydrous magnesium sulfate, filtered, and concentrated under reduced pressure using a rotary evaporator to give 29.2 g (82%) of crude (Z)-4-(trimethysilyl)-3-butenyl 4-methylbenzenesulfonate as a light yellow oil (Note 6).

B.   *Preparation of N-(4-methoxyphenyl)-(Z)-4-(trimethylsilyl)-3-butenamine.* A 250-mL, three-necked, round-bottomed flask is equipped with a magnetic stirring bar, a 250-mL addition funnel, and a gas inlet tube. The flask is flushed with argon or nitrogen and charged with 41.5 g (0.337 mol) of 4-methoxyaniline (Note 7) and then heated to 65°C. The stirring melt is degassed (Note 8), 20.2 g (67.8 mmol) of (Z)-4-(trimethylsilyl)-3-butenyl 4-methylbenzenesulfonate is added over 15 min, and the resulting solution is maintained at 65°C for 3 hr. The reaction product is allowed to cool to ca. 50°C and is then transferred to a 500-mL separatory funnel using 250 mL of chloroform. The chloroform solution is washed with two 100-mL portions of 1 M sodium hydroxide and the combined aqueous phases are extracted with 500 mL of chloroform. The combined organic phases are dried over anhydrous sodium sulfate, filtered, and concentrated under reduced pressure using a rotary evaporator. The crude residue is distilled through a 17-cm Vigreux column and excess 4-methoxyaniline is collected in the first fraction, bp 80-86°C (0.25 mm) (Note 9). Vacuum distillation is continued to give 10.6 g (63% yield) of

189

N-(4-methoxyphenyl)-(Z)-4-(trimethylsilyl)-3-butenamine, bp 125-128°C (0.25 mm), as a pale yellow oil (Note 10).

C. *Preparation of 1-(4-methoxyphenyl)-1,2,5,6-tetrahydropyridine.* An oven-dried, 250-mL, two-necked round-bottomed flask is equipped with a magnetic stirring bar, reflux condenser, and an argon or nitrogen inlet. The flask is flushed with argon or nitrogen, charged with 6.62 g (26.5 mmol) of N-(4-methoxyphenyl)-(Z)-4-(trimethylsilyl)-3-butenamine, 7.45 g (260 mmol) of paraformaldehyde (Note 11), 4.8 g (25 mmol) of p-toluenesulfonic acid monohydrate (Note 12) and 100 mL of acetonitrile (Note 13). The reaction mixture is degassed (Note 8) and heated at reflux for 1 hr (Note 14). The reaction mixture is cooled to room temperature and the excess paraformaldehyde is removed by vacuum filtration. The reaction vessel is washed with two 25-mL portions of dichloromethane and the washings are clarified by filtration. The combined organic phases are concentrated under reduced pressure using a rotary evaporator and the resulting solid residue is dissolved in dichloromethane and transferred to a 500-mL separatory funnel. The organic phase is washed with two 100-mL portions of 4 M sodium hydroxide and the aqueous washings are extracted with 50 mL of dichloromethane. The combined organic phases are then washed with 100 mL of water, dried over anhydrous potassium carbonate, filtered, and concentrated under reduced pressure using a rotary evaporator. The crude residue is dissolved in 9:1 hexane-ether and filtered through a 20-cm column (6-cm diameter) of silica gel. Evaporation of solvent gives 4.2 g (84% yield) of 1-(4-methoxyphenyl)-1,2,5,6-tetrahydropyridine as a white crystalline solid, mp 49-51°C (Notes 15, 16).

2. Notes

1. The trimethylsilyl butenol was prepared as described in the previous
Organic Syntheses procedure.

2. Pyridine is freshly distilled from calcium hydride under an argon
atmosphere.

3. p-Toluenesulfonyl chloride was purchased from Aldrich Chemical
Company, Inc. and was purified by dissolving 100 g in 100 mL of chloroform,
adding 1250 mL of hexane, filtering to remove insoluble impurities, and
concentrating the filtrate under reduced pressure.[2]

4. During this time, pyridinium hydrochloride precipitates from the
solution as white needles.

5. Caution must be exercised so that the ether layer does not become too
warm during this extraction.

6. The sample has the following spectral characteristics: IR (neat)
cm^{-1}: 1610, 1365, 1255, 1180. ^1H NMR ($CDCl_3$, 250 MHz) δ: 0.13 (s, 9 H,
SiC$\underline{H}_3$), 2.49-2.58 (m, 5 H), 4.14 (apparent t, 2 H, J = 6.9, C$\underline{H}_2$OR), 5.69 (d, 1
H, J = 14.1, R_3SiC$\underline{H}$=CH), 6.17 (overlapping dt, 1 H, J = 14.1, J = 7.2,
R_3SiCH=C$\underline{H}$), 7.40 (apparent d, 2 H, J = 7.8, aryl H), 7.85 (apparent d, 2 H, J
= 8.3, aryl H).   Gas chromatographic analysis using a 25-m 5% methylphenylsilicone column showed that this sample was >92% pure and contained several
unidentified impurities.

7. 4-Methoxyaniline (p-anisidine) was purchased from Aldrich Chemical
Company, Inc.

8. This is done by applying a mild vacuum to the reaction vessel and
then filling the vessel with argon or nitrogen. This operation was repeated
three times.

9. The condenser is not cooled and the collector tip is at times gently heated with a heat gun to prevent crystallization of 4-methoxyaniline in the distillation apparatus.

10. The product had the following spectral characteristics: IR (neat) cm^{-1}: 3390, 2950, 1608, 1246, 1040, 838; ^{1}H NMR (CDCl$_3$, 250 MHz) δ: 0.14 (s, 9 H, SiCH$_3$), 2.42-2.51 (apparent q, 2 H, J = 7, =CHCH$_2$), 3.15 (t, 2 H, J = 6.8, CH$_2$NR), 3.76 (s, 3 H, ArOCH$_3$), 5.67 (d, 1 H, J = 14.1, R$_3$SiCH=CH), 6.32 (overlapping dt, 1 H, J = 14.1 and 7.3, R$_3$SiCH=CH), 6.59 (apparent d, 2 H, J = 9.0, aryl H), 6.76 (apparent d, 2 H, J = 9.0, aryl H). High resolution mass spectrum (EI, 70 eV) 249.1548 (Calcd for C$_{14}$H$_{23}$NOSi: 249.1549). Gas chromatographic analysis using a 25-m 5% methylphenylsilicone capillary column showed that this sample was >95% pure. Two impurities of similar retention time, presumed to be the (E)-stereoisomer and the corresponding alkane, comprise from 1-3% of the product mixture depending on the run, while a third, longer retention time impurity, the corresponding tertiary amine, comprises 2% of the product mixture.

11. Paraformaldehyde was purchased from Alpha Products, Morton/Thiokol Inc.

12. p-Toluenesulfonic acid monohydrate was purchased from Aldrich Chemical Company, Inc. and is suitable for use after storage for 24 hr in a vacuum desiccator over phosphorus pentoxide..

13. Acetonitrile was purchased from Mallinkrodt, Inc.

14. During this time paraformaldehyde can be seen forming on the inside of the reflux condenser.

15. The sample thus obtained is 94-97% pure by capillary GC analysis using a 25-m 5% methylphenylsilicone capillary column. This material gave the following elemental analysis: Anal. Calcd for C$_{12}$H$_{15}$NO: C, 76.15; H, 7.99; N, 7.40. Found: C, 75.49; H, 8.15; N, 7.31.

16. A purer sample may be obtained by vacuum sublimation at 60°C (0.3 mm). The material shows the following spectral characteristics: IR (KBr) cm^{-1}: 2831, 1514, 1249, 1210, 1190, 1035, 815; ^{1}H NMR (250 MHz) δ: 2.4-2.7 (m, 2 H), 3.27 (t, 2 H, J = 5.6), 3.58-3.65 (m, 2 H), 3.80 (s, 3 H, OCH$_3$), 5.7-5.9 (m, 2 H, RC$\underline{H}$=C$\underline{H}$R), 6.85-6.95 (m, 4 H, aryl $\underline{H}$). Gas chromatographic analysis using a 25-m 5% methylphenylsilicone column showed that this material was 98% pure and was contaminated with 1.8% of the starting secondary amine and 0.3% of the corresponding tertiary acyclic amine. This material melts at 50-52°C and gave the following elemental analysis: Anal. Calcd for C$_{12}$H$_{15}$NO: C, 76.15; H, 7.99; N, 7.40. Found: C, 76.18; H, 8.00; N, 7.40. The oxalate salt melts at 134-135°C and gave the following elemental analysis: Anal. Calcd for C$_{14}$H$_{17}$NO$_5$: C, 60.21; H, 6.09; N, 5.01. Found: C, 60.09; H, 6.16; N, 4.98.

3. Discussion

A variety of 1,2,5,6-tetrahydropyridines can be prepared by the reaction of (Z)-4-(trimethylsilyl)-3-butenamines with aldehydes.[3,4,5] Representative examples are summarized in Table I. Cyclizations with paraformaldehyde occur readily in refluxing acetonitrile, while cyclizations with other aldehydes require higher temperatures. Tetrahydropyridines with substituents at atoms -1, -2, -3, and -4 have been regioselectively prepared in this way. In no case was any trace of a regioisomeric tetrahydropyridine detected.

The 1,2,5,6-tetrahydropyridine ring is found in several natural products and numerous pharmacologically active materials.[5] This ring system is most commonly constructed by reduction of the corresponding pyridinium salt or from 4-piperidone precursors.[5a] The cyclization approach reported here has the advantage of complete regiocontrol of the double-bond position. Moreover this approach is of particular value for the synthesis of 1-aryl-substituted tetrahydropyridines that are difficult to access, since they are not generally available from pyridine precursors.

Iminium ion-vinylsilane cyclizations closely related to the one described here have been used to prepare indolizidine alkaloids of the pumiliotoxin A[6] and elaeokanine[3] families, indole alkaloids,[7] amaryllidaceae alkaloids,[8] and the antibiotic (+)-streptazolin.[9] The ability of the silicon substituent to control the position, and in some cases stereochemistry, of the unsaturation in the product heterocycle was a key feature of each of these syntheses.

Alternative methods for preparing 1-(4-methoxyphenyl)-1,2,5,6-tetrahydropyridine have not been reported.

TABLE I

PREPARATION OF SUBSTITUTED 1,2,5,6-TETRAHYDROPYRIDINES[3,4]

R^1	R^2	R^3	R^4	Cyclization Step Conditions Temp., °C; Time, hr	Yield, %
C_3H_7	H	H	H	80; 1.5	61
4-Methoxybenzyl	H	H	H	80; 1.5	91
Cyclohexyl	H	H	H	110; 10	54[a]
Ph	H	H	H	80; 0.7	61
4-Methoxyphenyl	H	H	H	80; 1	84
C_3H_7	C_6H_{13}	H	H	120; 48	54
4-Methoxybenzyl	C_6H_{13}	H	H	120; 72	64
Ph	C_6H_{13}	H	H	80; 3	68
Iso-C_4H_9	H	H	CH_3	80; 2	66
Iso-C_4H_9	H	H	Ph	80; 2	83
C_3H_7	H	$SiMe_3$	H	80; 1.2	82

[a]In this case the reaction of a cyanomethyl tertiary amine with silver trifluoroacetate in chloroform was used to initiate the cyclization instead of the reaction of an aldehyde with a secondary amine salt.

1. Department of Chemistry, University of California, Irvine, CA 92717.

2. Fieser, L. F.; Fieser, M. "Reagents for Organic Syntheses", John Wiley and Sons: New York, 1967; Vol. I., p. 1180.

3. Overman, L. E.; Malone, T. C.; Meier, G. P. *J. Am. Chem. Soc.* **1983**, *105*, 6993; Flann, C. J.; Malone, T. C.; Overman, L. E. *J. Am. Chem. Soc.* **1987**, *109*, 6097.

4. Overman, L. E. *Lect. Heterocycl. Chem.* **1985**, *8*, 59; Blumenkopf, T. A.; Overman, L. E. *Chem. Rev.* **1986**, *86*, 857.

5. (a) Fowler, F. W. in "Comprehensive Heterocyclic Chemistry"; Katritzky, A. R.; Rees, C. W., Eds.; Pergammon Press: London, 1984; Vol. 2, Part 2A, pp. 365-394; (b) Coutts, R. T.; Scott, J. R. *Can. J. Pharm. Sci.* **1971**, *6*, 78.

6. Overman, L. E.; Bell, K. L. *J. Am. Chem. Soc.* **1981**, *103*, 1851; Overman, L. E.; Bell, K. L.; Ito, F. *J. Am. Chem. Soc.* **1984**, *106*, 4192; Overman, L. E.; Lin, N.-H. *J. Org. Chem.* **1985**, *50*, 3669.

7. Overman, L. E.; Malone, T. C. *J. Org. Chem.* **1982**, *47*, 5297.

8. Overman, L. E.; Burk, R. M. *Tetrahedron Lett.* **1984**, *25*, 5737.

9. Flann, C. J.; Overman, L. E. *J. Am. Chem. Soc.* **1987**, *109*, 6115.

Appendix

Chemical Abstracts Nomenclature (Collective Index Number);

(Registry Number)

(Z)-4-(Trimethylsilyl)-3-butenyl 4-methylbenzenesulfonate: 3-Buten-1-ol,
4-(trimethylsilyl)-, 4-methylbenzenesulfonate, (Z)- (11); (87682-62-0)

(Z)-4-(Trimethylsilyl)-3-buten-1-ol: 3-Buten-1-ol, 4-(trimethylsilyl)-,
(Z)- (11); (87682-77-7)

p-Toluenesulfonyl chloride (8); (Benzenesulfonyl chloride, 4-methyl- (9);
(98-59-9)

4-Methoxyaniline: p-Anisidine (8); Benzenamine, 4-methoxy- (9); (104-94-9)

Paraformaldehyde: Poly(oxymethylene) (8,9); (9002-81-7)

p-Toluenesulfonic acid monohydrate (8); Benzenesulfonic acid, 4-methyl-,
monohydrate (9); (6192-52-5)

A.

B.

C.

Submitted by Richard T. Taylor,[1] Michael W. Pelter,[1] and Leo A. Paquette.[2]
Checked by Katsunori Nagai and Ryoji Noyori.

1. Procedure

All apparatus for Steps A and B should be dried overnight in an oven.

A. *9,10-Dihydrofulvalene.* A 5-L, three-necked, round-bottomed flask is
fitted, while hot, with a Hirschberg stirrer, gas inlet, and stopper (Note
1). The assembled apparatus is flame-dried and allowed to cool to room
temperature under a stream of dry, oxygen-free argon (Note 2). The stopper is

replaced with a powder funnel and, under a sweep of positive argon, 100 g (4.17 mol) of dry sodium hydride (Note 3) is added followed by 2.0 L of dry tetrahydrofuran (Note 4). The powder funnel is replaced by a 500-mL, pressure-equalizing, jacketed addition funnel which is flushed with argon, then stoppered.

The stirred sodium hydride suspension is cooled by an external ice-water bath and the jacket of the addition funnel is cooled in a dry ice-isopropyl alcohol bath. Into the addition funnel is introduced 275 g (4.16 mol) of neat, freshly distilled cyclopentadiene (Note 5). The cyclopentadiene is added rapidly, dropwise over 30-40 min to the stirred slurry (*Caution: Avoid excess foaming*) (Note 6). After the addition is complete, the cooling bath is removed and the solution is stirred for 1 hr at room temperature.

The jacketed addition funnel is removed and 1.5 g of cuprous bromide-dimethyl sulfide complex (Note 7) is added through a powder funnel. A 500-mL, pressure-equalizing addition funnel (long-tipped) is attached to the flask and flushed with argon. As the anion solution is cooled in a dry ice-isopropyl alcohol bath, a solution of 530 g (2.08 mol) of sublimed iodine in 500 mL of anhydrous tetrahydrofuran is placed in the addition funnel. This solution is added dropwise to the cooled slurry over approximately 90 min (Note 8). The solution is stirred for about 15 min at low temperature.

B. *Diels-Alder reaction.* A 500-mL, pressure-equalizing addition funnel containing 330 g (2.32 mol) of dimethyl acetylenedicarboxylate (Note 9) is placed in the flask and the ester is added rapidly dropwise over 10 min. The solution is stirred for 30 min, the cooling bath is removed, and stirring is maintained for 4 hr (Note 10).

The reaction solution is filtered through a Celite pad (approximately 5 cm thick on a 32-cm Büchner funnel), and the solid is washed repeatedly with tetrahydrofuran (1.5 L). The combined filtrates are concentrated under reduced pressure at a temperature not above 30°C. To the concentrate is added 1.5 L of ether. The solution is stirred for 15 min, again filtered through Celite, and concentrated at 30°C (Note 11).

C. Hydrolysis. Into a three-necked, 5-L, round-bottomed flask equipped with a mechanical stirrer, thermometer, and 500-mL addition funnel with gas inlet is placed the above concentrate and 2 L of methanol. The solution is cooled to -5° to 0°C by means of an ice-salt bath. A precooled (0°C) solution containing 220 g of 87.5% potassium hydroxide in 400 mL of water is added dropwise at such a rate as to keep the reaction temperature below 10°C. The reaction mixture is stirred for an additional 2 hr at 0°C and for 1 hr at room temperature prior to the addition of 100 mL of glacial acetic acid. Solid sodium carbonate is added to bring the pH to 8 and the solution is filtered through Celite. Concentration of the filtrate at 35°C and reduced pressure affords about 1 L of a dark liquid. The liquid is diluted with 2 L of water and extracted with petroleum ether (6 x 600 mL). The combined extracts are washed with aqueous sodium thiosulfate solution (800 mL) and dried (magnesium sulfate). Concentration at 30°C affords a clear red liquid (occasionally a yellow solid) which is almost pure internal diester (Note 12).

The diester is dissolved in 130 mL of methanol, placed in a 1-L, one-necked flask equipped with a magnetic stirrer bar and reflux condenser, and treated with a solution containing 35 g of potassium hydroxide in 130 mL of water. The mixture is stirred at reflux temperature for 1 hr. Methanol is removed under reduced pressure and 250 mL of water is added. Heating is continued for another 5 hr. After the solution is cooled, 5 g of activated charcoal is added and the mixture is stirred for 8 hr at room temperature. Filtration through Celite is followed by cooling of the stirred filtrate in an ice bath with acidification to pH 1 (dropwise addition of concentrated hydrochloric acid). The tan solid is isolated by filtration and dried under vacuum at room temperature. On the average, the yield is 52-55 g (10-11%) but can vary from 42-68 g (8-13% yield) (Note 13).

2. Notes

1. A paste made from a 1:1 mixture of mineral oil and silicone grease is used to lubricate the stirrer.

2. Prepurified argon (Linde) can be used with no further treatment.

3. Dry sodium hydride is available from Aldrich Chemical Company, Inc., as a fine powder. In multiple runs, it is most convenient to weigh the bulk reagent into 100-g (one reaction) lots. Extreme care should be used in handling this moisture-sensitive, flammable solid.

4. Tetrahydrofuran is distilled from calcium hydride and then from sodium-benzophenone immediately prior to use.

5. This amount of cyclopentadiene can be prepared in 2-3 hr using any of a variety of procedures.[3-5]

6. The checkers transferred cyclopentadiene by using a stainless steel cannula from a cooled (dry ice-methanol), 500-mL, round-bottomed flask to the reaction vessel.

7. The complex was purchased from Aldrich Chemical Company, Inc.

8. A bright emerald green color (usually) develops as the addition proceeds. If the iodine is impure, a brown color develops with no decrease in yield.

9.   While dimethyl acetylenedicarboxylate is available commercially, it is easily prepared by the procedure of Huntress, Lesslie, and Bornstein.[6] Care should be taken with this compound as it is a severe lachrymator and vesicant.

10.   A tan to white precipitate of sodium iodide forms and a gentle exotherm is observed.

11.   This concentrate (a dark red oil) may be stored in a refrigerator if time does not permit further work.

12.   While the Diels-Alder reaction affords a wide variety of products, all of the esters formed hydrolyze faster than the desired internal adduct. The above hydrolysis removes all byproducts through base extraction.   The internal diester has spectral properties as follows: ^{1}H NMR (CDCl$_3$) δ: 2.50 (tt, 2 H, J = 2.0, 4.3), 3.30 (dd, 4 H, J = 2.0, 4.3), 3.59 (s, 6 H), 6.07 (t, 4 H, J = 2.0); ^{13}C NMR (CDCl$_3$) δ: 51.5, 58.8, 64.4, 69.5, 132.7, 172.7.

13.   Spectral properties of the diacid are as follows: ^{1}H NMR (DMSO-d$_6$) δ: 2.36 (tt, 2 H, J = 1.5, 4.1), 3.21 (dd, 4 H, J = 1.5, 4.1), 5.95 (t, 4 H, J = 1.7).

3. Discussion

The present procedure is a modification of the method previously reported.[7] While the overall yield is similar, the method described here is simpler in that it avoids a cumbersome transfer of the sodium cyclopentadienide solution.

The first step leading to 9,10-dihydrofulvalene is adapted from the earlier work of Matzner.[8] The utility of this thermally labile hydrocarbon ranges from its ability to engage in multiple [4+2] cycloadditions[7,9] to its capacity for bonding to a pair of metal atoms.[10]

The product diacid has served as starting material for the synthesis of tetracyclo[7.2.1.0^{4,11}.0^{6,10}]dodeca-2,7-diene-5,12-dione,[11] C$_{16}$-hexaquinacene,[12] (C$_s$)-C$_{17}$-heptaquinane derivatives,[13] the parent dodecahedrane molecule,[14] and a number of substituted dodecahedranes.[15]

1. Department of Chemistry, Miami University, Oxford, OH 45056.

2. Department of Chemistry, The Ohio State University, Columbus, OH 43210.

3. Moffett, R. B. *Org. Synth., Collect. Vol. 4* **1963**, 238.

4. Korach, M.; Nielsen, D. R.; Rideout, W. H. *Org. Synth., Collect. Vol. 5* **1973**, 414.

5. Magnusson, G. *J. Org. Chem.* **1985**, *50*, 1998.

6. Huntress, E. H.; Lesslie, T. E.; Bornstein, J. *Org. Synth., Collect. Vol. 4* **1963**, 329.

7. (a) Paquette, L. A.; Wyvratt, M. J. *J. Am. Chem. Soc.* **1974**, *96*, 4671; (b) Paquette, L. A.; Wyvratt, M. J.; Berk, H. C.; Moerck, R. E. *J. Am. Chem. Soc.* **1978**, *100*, 5845.

8. Matzner, E. A. Ph.D. Thesis, Yale University, New Haven, CT, 1958.

9. Wyvratt, M. J.; Paquette, L. A. *Tetrahedron Lett.* **1974**, 2433.

10. (a) Vollhardt, K. P. C.; Weidman, T. W. *J. Am. Chem. Soc.* **1983**, *105*, 1676; Vollhardt, K. P. C.; Weidman, T. W. *Organometallics* **1984**, *3*, 82; (b) Drage, J. S.; Tilset, M.; Vollhardt, K. P. C.; Weidman, T. W. *Organometallics* **1984**, *3*, 812; (c) Drage, J. S.; Vollhardt, K. P. C. *Organometallics* **1985**, *4*, 191; Drage, J. S.; Vollhardt, K. P. C. *Organometallics* **1986**, *5*, 280; (d) Tilset, M.; Vollhardt, K. P. C. *Organometallics* **1985**, *4*, 2230; (e) Boese, R.; Tolman, W. B.; Vollhardt, K. P. C. *Organometallics* **1986**, *5*, 582.

11. (a) Paquette, L. A.; Nakamura, K.; Fischer, J. W. *Tetrahedron Lett*. **1985**, *26*, 4051; (b) Paquette, L. A.; Nakamura, K.; Engel, P. *Chem. Ber*. **1986**, *119*, 3782.

12. (a) Paquette, L. A.; Snow, R. A.; Muthard, J. L.; Cynkowski, T. *J. Am. Chem. Soc*. **1978**, *100*, 1600; (b) Paquette, L. A.; Snow, R. A.; Muthard, J. L.; Cynkowski, T. *J. Am. Chem. Soc*. **1979**, *101*, 6991; (c) Osborn, M. E.; Kuroda, S.; Muthard, J. L.; Kramer, J. D.; Engel, P.; Paquette, L. A. *J. Org. Chem*. **1981**, *46*, 3379; (d) Christoph, G. G.; Muthard, J. L.; Paquette, L. A.; Bohm, M. C.; Gleiter, R. *J. Am. Chem. Soc*. **1978**, *100*, 7782.

13. (a) Sobczak, R. L.; Osborn, M. E.; Paquette, L. A. *J. Org. Chem*. **1979**, *44*, 4886; (b) Osborn, M. E.; Pegues, J. F.; Paquette, L. A. *J. Org. Chem*. **1980**, *45*, 167.

14. (a) Ternansky, R. J.; Balogh, D. W.; Paquette, L. A. *J. Am. Chem. Soc*. **1982**, *104*, 4503; (b) Paquette, L. A.; Ternansky, R. J.; Balogh, D. W.; Kentgen, G. *J. Am. Chem. Soc*. **1983**, *105*, 5446; (c) Gallucci, J. C.; Doecke, C. W.; Paquette, L. A. *J. Am. Chem. Soc*. **1986**, *108*, 1343.

15. (a) Paquette, L. A.; Balogh, D. W.; Usha, R.; Kountz, D.; Christoph, G. G. *Science* **1981**, *211*, 575; (b) Paquette, L. A.; Balogh, D. W. *J. Am. Chem. Soc*. **1982**, *104*, 774; (c) Christoph, G. G.; Engel, P.; Usha, R.; Balogh, D. W.; Paquette, L. A. *J. Am. Chem. Soc*. **1982**, *104*, 784; (d) Paquette, L. A.; Ternansky, R. J.; Balogh, D. W. *J. Am. Chem. Soc*. **1982**, *104*, 4502; (e) Paquette, L. A.; Balogh, D. W.; Ternansky, R. J.; Begley, W. J.; Banwell, M. G. *J. Org. Chem*. **1983**, *48*, 3282; (f) Paquette, L. A.; Ternansky, R. J.; Balogh, D. W.; Taylor, W. J. *J. Am. Chem. Soc*. **1983**, *105*, 5441; (g) Paquette, L. A.; Miyahara, Y.; Doecke, C. W. *J. Am. Chem. Soc*. **1986**, *108*, 1716; (h) Paquette, L. A.; Miyahara, Y. *J. Org. Chem*. **1987**, *52*, 1265.

Appendix

Chemical Abstracts Nomenclature (Collective Index Number);

(Registry Number)

3,3a,3b,4,6a,7a-Hexahydro-3,4,7-metheno-7H-cyclopenta[a]pentalene-7,8-
dicarboxylic acid: 3,4,7-Metheno-7H-cyclopenta[a]pentalene-7,8-dicarboxylic
acid, 3,3a,3b,4,6a,7a-hexahydro- (10); (61206-25-5)

9,10-Dihydrofulvalene: Bi-2,4-cyclopentadien-1-yl (8,9); (21423-86-9)

Cyclopentadiene: 1,3-Cyclopentadiene (8,9); (542-92-7)

Dimethyl acetylenedicarboxylate: Acetylenedicarboxylic acid, dimethyl ester
(8); 2-Butynedioic acid, dimethyl ester (9); (762-42-5)

Dimethyl 3,3a,3b,4,6a,7a-hexahydro-3,4,7-metheno-7H-cyclopenta[a]pentalene-
7,8-dicarboxylate: 3,4,7-Metheno-7H-cyclopenta[a]pentalene-7,8-dicarboxylic
acid, 3,3a,3b,4,6a,7a-hexahydro-, dimethyl ester (9); (53282-97-6)

IMMONIUM ION-BASED DIELS-ALDER REACTIONS:

N-BENZYL-2-AZANORBORNENE

(2-Azabicyclo[2.2.1]hept-5-ene, 2-(phenylmethyl)-)

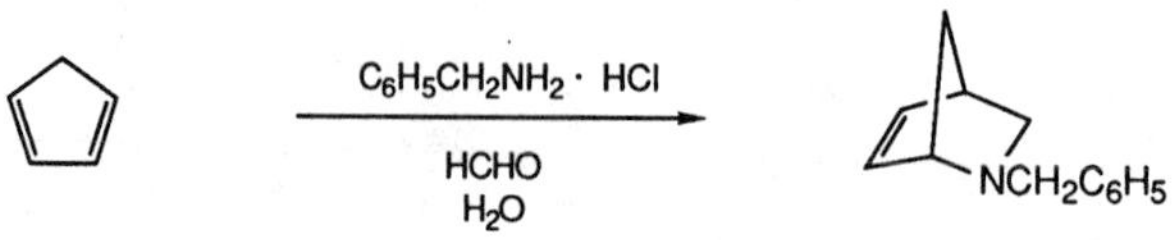

Submitted by Paul A. Grieco and Scott D. Larsen.[1]

Checked by V. Ramamurthy and Bruce E. Smart.

1. Procedure

A 100-mL round-bottomed flask equipped with a Teflon-coated magnetic stirring bar is charged with 24 mL of de-ionized water and 8.6 g (60.0 mmol) of benzylamine hydrochloride (Note 1). To the above homogeneous solution is added 6.3 mL (84 mmol) of 37% aqueous formaldehyde solution (Note 2) followed by 9.9 mL (120 mmol) of freshly prepared cyclopentadiene (Note 3). The flask is stoppered tightly (Note 4) and stirred vigorously at ambient temperature. After 4 hr, the reaction mixture is poured into 50 mL of water and washed with ether-hexane, 1:1 (2 x 40 mL). The aqueous phase is made basic by the addition of 4.0 g of solid potassium hydroxide and extracted with ether (3 x 60 mL). The combined ether extracts are dried over anhydrous magnesium sulfate and filtered. The solvent is removed under reduced pressure (15-20 mm) to give 11.2 g (100%) of N-benzyl-2-azanorbornene as a very pale yellow oil (Note 5). The crude product is distilled at 80-85°C (0.05 mm) (Note 6) through a short path apparatus to provide 10.1-10.2 g (91-92%) of pure product (Note 7) as a colorless oil (Note 8).

2. Notes

1. Benzylamine hydrochloride is commercially available from Aldrich Chemical Company, Inc.

2. Aqueous formaldehyde solution (37%) is commercially available from Mallinckrodt Inc.

3. Cyclopentadiene is prepared by heating commercial dicyclopentadiene (available from Aldrich Chemical Company, Inc.) at 160°C in a distillation apparatus. Cyclopentadiene distills smoothly at 39-45°C.[2]

4. The heterogeneous reaction mixture is stoppered tightly to avoid loss of cyclopentadiene.

5. This crude material is essentially pure product contaminated by trace amounts of ether. N-Benzyl-2-azanorbornene has the following spectrum: ^{1}H NMR (300 MHz, $CDCl_3$) δ: 1.42 (dm, 1 H, J = 8), 1.52 (dd, 1 H, J = 2, 8.5), 1.64 (dm, 1 H, J = 8), 2.94 (bs, 1 H), 3.18 (dd, 2 H, J = 3, 8.5), 3.34, 3.58 (AB, 2 H, J = 13), 3.83 (m, 1 H), 6.09 (dd, 1 H, J = 2, 6), 6.38 (ddd, 1 H, J = 2, 3, 6), 7.2-7.4 (m, 5 H).

6. Attempted distillation at 15-20 mm resulted in extensive decomposition.

7. The submitters obtained 10.8 g (97%) of analytically pure product, bp 80-85°C (0.05 mm). Anal. Calcd for $C_{13}H_{15}N$: C, 84.28; H, 8.16; N, 7.56. Found: C, 84.68; H, 8.36; N, 7.59.

8. Upon prolonged standing in air at room temperature discoloration of the product accompanied by slow evolution of cyclopentadiene takes place.

3. Discussion

Simple unactivated immonium salts generated in situ from formaldehyde and primary alkyl amines undergo a facile aza Diels-Alder reaction with cyclopentadiene at ambient temperature[3] to afford novel N-alkylated 2-azanorbornenes. The procedure described above is general and can be applied to a number of primary alkyl amines. Yields of N-alkyl substituted 2-azanorbornenes are good to excellent. Use of ammonium chloride and formaldehyde in the above reaction produces 2-azanorbornene in modest (40-50%) yield. 2-Azanorbornene (3) has been previously prepared[4] by reaction of cyclopentadiene with chlorosulfonyl isocyanate which provides a single N-chlorosulfonyl β-lactam (1). Exposure of 1 to an aqueous solution of sodium sulfite gives rise (25-30%) to 2-azanorbornen-3-one (2) which upon reduction with lithium aluminum hydride affords (ca. 80%) 2-azanorbornene (3).

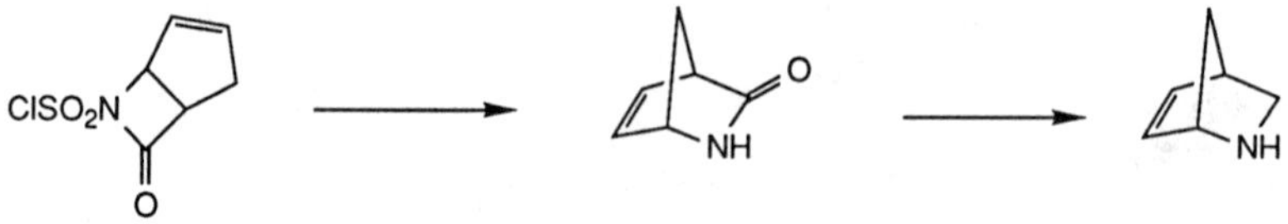

1. Department of Chemistry, Indiana University, Bloomington, IN 47405.

2. Moffett, R. B. *Org. Synth., Collect. Vol. 4*, **1963**, 238.

3. Larsen, S. D.; Grieco, P. A. *J. Am. Chem. Soc.* **1985**, *107*, 1768.

4. Malpass, J. R.; Tweddle, N. J. *J. Chem. Soc., Perkin Trans. I*, **1977**, 874.

Appendix

Chemical Abstracts Nomenclature (Collective Index Number); (Registry Number)

Benzylamine hydrochloride (8); Benzenemethanamine, hydrochloride (9); (3287-99-8)

Formaldehyde (8,9); (50-00-0)

Cyclopentadiene: 1,3-Cyclopentadiene (8,9); (542-92-7)

THE CARROLL REARRANGEMENT: 5-DODECEN-2-ONE

(5-Dodecen-2-one, (E)-)

Submitted by Stephen R. Wilson and Corinne E. Augelli.[1]

Checked by T. R. Vedananda and James D. White.

1. Procedure

A. *(1-Ethenyl)heptanyl 3-ketobutanoate.* 3-Hydroxy-1-nonene (Note 1, 7.5 g 0.053 mol) is stirred in 250 mL of anhydrous ether (Note 2) in a 500-mL, three-necked, round-bottomed flask, fitted with a thermometer, nitrogen/mineral oil bubbler, Teflon-covered magnetic stirring bar, and a rubber septum. A constant flow of nitrogen is maintained throughout the reaction. To this clear homogeneous solution, diketene (Note 3, 5.04 g, 0.060 mol) is added in one portion by syringe, followed by 4-dimethylaminopyridine

(DMAP) (0.591 g, 0.0049 mol, Note 4). A slightly exothermic reaction (to 32°C) is observed. After 15 min, the reaction is complete (Note 5). The reaction mixture is quenched with 100 mL of 0.1% sodium hydroxide solution and 100 mL of anhydrous ether. The layers are separated and the organic phase is washed once with 50 mL of 0.1% sodium hydroxide solution and once with brine. Drying over magnesium sulfate and concentration under reduced pressure affords 11.3 g (0.05 mol, 94%) of a pale yellow oil. The β-keto ester is of high purity, as shown by TLC, GLC, and spectral analyses (Note 6); however, distillation (bp 113-114°C/1.4 mm) leads to a colorless oil.

B. *3-Carboxy-5-dodecen-2-one.* A solution of lithium diisopropylamide (LDA, Note 7) in 150 mL of tetrahydrofuran (Note 8) is prepared from 0.199 mol (20.12 g) of diisopropylamine (Note 9) and 0.181 mol of butyllithium (Notes 10,11). The solution is cooled to -78°C (acetone/dry ice) and a solution of 10.0 g (0.044 mol) of (1-ethenyl)heptanyl 3-ketobutanoate in 50 mL of tetrahydrofuran (Note 8) is added via a 125-mL, pressure-equalizing dropping funnel at -78°C over a 15-min period. After the addition is complete, the reaction mixture is stirred at -78°C for 45 min and is then allowed to warm gradually to room temperature. When the reaction mixture finally reaches 25°C (about 2 hr), it has a deep red color. After the mixture is stirred for 18 hr at room temperature, the reaction is complete (Note 12). To this mixture, 100 mL of water and 100 mL of pentane are added in portions with stirring, maintaining the temperature below 25°C with an ice bath. As the deep red reaction mixture is quenched, an orange heterogeneous mixture results. The layers are separated and the pentane layer is extracted two times with 50 mL of 0.1% sodium hydroxide solution (Note 13). All aqueous layers are combined in an 800-mL beaker equipped with a Teflon-covered magnetic stirring bar. A 100-mL aliquot of pentane is added to this aqueous mixture which is stirred vigorously, and then 100 mL of 10% hydrochloric acid solution is added in 10-mL portions until pH 2 is reached (Note 14). The heterogeneous solution is poured into a 1-L separatory funnel and the layers are quickly separated. The aqueous layer is extracted three times more, each time with 50 mL of pentane. The combined organic layers (ca. 250-300 mL) are dried (MgSO$_4$) and evaporated at reduced pressure without heating to give 9.6 g (0.04 mol, 97%) of a red-orange oil. This carboxylic acid is of high purity as shown by TLC and spectral analyses (Note 15).

  C. *5-Dodecen-2-one*. The 3-carboxy-5-dodecen-2-one (9.0 g, 0.040 mol) is stirred in 150 mL of carbon tetrachloride (Note 16) in a 500-mL, three-necked, round-bottomed flask which is fitted with a reflux condenser, thermometer, Teflon-covered magnetic stirrer, and a ground-glass stopper. After the orange-yellow solution is heated at reflux for 1 hr, TLC analysis shows the reaction to be complete (Note 17). The reaction mixture is concentrated under reduced pressure with warming to afford 7.1 g (0.039 mol, 98%) of a red-orange oil. The product is sufficiently pure for most purposes. It may be purified by vacuum distillation at 105-107°C/2.7 mm (5.2 g, 0.03 mol, 71%) which yields a pale yellow oil (Note 18).

2. Notes

1.   3-Hydroxy-1-nonene was prepared by the following procedure: 0.219 mol (25.0 g) of heptaldehyde (Eastman Kodak Co.), distilled prior to use, was stirred in 450 mL of anhydrous ether (Note 2) under a nitrogen atmosphere. The solution was cooled to 0°C with an ice bath and 0.260 mol (260.4 mL) of vinylmagnesium bromide (1.0 M solution in tetrahydrofuran, Aldrich Chemical Company, Inc., 1.2 equiv) was then added dropwise over a 0.5-hr period. The

reaction mixture was allowed to warm gradually to room temperature and was stirred for 0.5 hr. The reaction mixture was added, in portions with stirring, to 400 mL of a saturated ammonium chloride solution, maintaining the temperature below 25°C with an ice bath. This quenched reaction mixture was stirred for 15 min. The layers were separated and the aqueous layer was extracted two times, each time with 200 mL of ether. The combined organic layers were extracted with 200 mL of brine, dried (MgSO$_4$) and concentrated under reduced pressure with warming to afford a yellow oil in quantitative yield. The allylic alcohol was distilled (bp 185°C/760 mm, 22.5 g 0.158 mol, 73%) before use. Silica gel TLC showed one spot: R$_f$ = 0.34 (20% ethyl acetate/ligroin). GLC: Retention time, 4.56 min; Program: 40°C/1 min; 20°C/1 min to 320°C; 2% OV-101, 0.2% Carbowax on Chromosorb. This program and column were used throughout the entire sequence of reactions. IR (neat) cm^{-1}: 3610 (OH); 3000 (C-H, alkenes); 1650 (C=C); 1000 (C-O). Mass spectrum: m/e 57 (100% M$^+$ - CH$_2$(CH$_2$)$_4$CH$_3$), 85 (7.7% M$^+$ - CHOHCH=CH$_2$), 113 (2.5% M$^+$ - CH$_2$CH$_3$). ^{1}H NMR (CDCl$_3$) δ: 0.91 (t, 1 H, J = 6.9); 1.54-1.31 (m, 11 H); 4.12 (q, 1 H, J = 6.3); 5.27-5.11 (m, 1 H); 5.95-5.84 (m, 1 H). ^{13}C NMR (CDCl$_3$) δ: 14.0, 22.6, 25.4, 29.3, 31.9, 37.1, 72.9, 114.0, 141.5.

2. Anhydrous ether (Fisher Scientific Company) was used without further drying .

3. Diketene (Aldrich Chemical Company, Inc.) was distilled immediately prior to use.

4. 4-Dimethylaminopyridine (DMAP) was purchased from Aldrich Chemical Company, Inc. A catalytic amount of 4-dimethylaminopyridine is necessary for this reaction to proceed.

5. Reaction progress can be monitored by GLC analysis of the disappearance of diketene.

6.   Silica gel TLC shows one spot at R_f = 0.45 (20% ethyl acetate/ligroin); GLC shows >95% purity of the β-keto ester. The β-keto ester undergoes partial cleavage to the corresponding allylic alcohol under these GLC conditions: Retention time, 7.8 ((1-ethenyl)heptanyl 3-ketobutanoate); retention time, 4.88 (3-hydroxy-1-nonene).   Mass spectrum for the β-keto ester: m/e 43 (100% M^+ - $COCH_3$), 85 (46.5% $(CH_2)_5CH_3$), 141 (13.3% M^+ - $COCH_2COCH_3$).   Mass spectrum for allylic alcohol: m/e 57 (100% M^+ - $CH_2(CH_2)_4CH_3$), 85 (8.7% M^+ - $CHOHCH=CH_2$), 113 (2.8% M^+ - CH_2CH_3).   IR (neat) cm^{-1}:   3000 (C-H, alkenes); 1725 (C=O); 1650 (C=C). ^1H NMR ($CDCl_3$) δ: 0.89 (t, 3 H); 1.30-1.64 (m, 10 H); 2.3 (s, 3 H), 3.47 (s, 2 H); 5.19-5.31 (m, 3 H); 5.72-5.84 (m, 1 H).   Anal. Calcd. for $C_{13}H_{22}O_3$:   C,   68.99; H, 9.8. Found: C, 69.25; H, 10.06.

7.   Lithium diisopropylamide (LDA) was prepared by the method described in *Org. Synth.* **1985**, *64*, 68-72.

8.   Tetrahydrofuran was distilled from lithium aluminum hydride immediately prior to use.

9.   Diisopropylamine, purchased from Aldrich Chemical Company, Inc., was distilled immediately prior to use.

10.   Butyllithium, 2.5 M solution, in hexanes was purchased from Aldrich Chemical Company, Inc.   Butyllithium was titrated with diphenylacetic acid[2] before each use.

11.   An amount of 4.1 equiv of lithium diisopropylamide (LDA) is absolutely necessary for this reaction to go to completion.   An equilibrium exists between the formation of the second anion of the β-keto ester and the formation of lithium diisopropylamide from diisopropylamine.

12. Reaction progress can be followed most accurately by silica gel TLC and GLC analysis. In TLC analysis, one sees the disappearance of (1-ethenyl)-

heptanyl 3-ketobutanoate, R_f = 0.45 (20% ethyl acetate/ligroin) and the appearance of baseline material which is indicative of the corresponding carboxylic acid salt. In GLC analysis, an aliquot (3 drops of reaction mixture, 3 drops of ether, 3 drops of 10% hydrochloric acid) will show complete disappearance of peaks at R_t = 7.44 and 4.56 (which correspond to (1-ethenyl)heptanyl 3-ketobutanoate and cleavage of this β-keto ester under GLC conditions to the 3-hydroxy-1-nonene, respectively) and appearance of a peak at R_t = 7.04 which corresponds to 5-dodecen-2-one. The 3-carboxy-5-dodecen-2-one decarboxylates upon injection yielding the GLC spectrum of the ultimate product.

13. Sodium hydroxide (0.1%) is used to extract all of the carboxylate from the ether layer in the form of the sodium salt.

14. A slight exotherm was noted from 25°C to 32°C. Acidification is necessary to extract all of the desired carboxylate from the aqueous layer into the organic layer.

15. Silica gel TLC shows one major spot at the baseline (20% ethyl acetate/ligroin) with a very slight impurity (<5%) at R_f = 0.47. Mass spectrum of 5-dodecen-2-one: m/e 43 (100% $COCH_3$), 97 (6.1% M^+ - $CH_2(CH_2)_4CH_3$), 125 (1.7% M^+ - CH_2COCH_3). IR (neat) cm^1: 3000 (COOH); 1725 (COOH, RCOR). ^1H NMR ($CDCl_3$) δ: 0.90 (t, 3 H, J = 6.3); 1.28 (m, 8 H); 2.00 (m, 2 H); 2.31 (s, 3 H); 2.60 (t, 2 H, J = 6.9); 3.48-3.59 (m, 1 H); 5.27-5.44 (m, 1 H, trans J = 15.3); 5.49-5.61 (m, 1 H, trans J = 15.3).

16. Carbon tetrachloride was used as purchased from Fisher Scientific Company.

17. Reaction progress was followed most accurately by TLC analysis. The disappearance of the baseline material (i.e., 3-carboxy-5-dodecen-2-one) and appearance of the desired ketone at R_f = 0.33 (10% ethyl acetate/ligroin) indicates the completeness of the reaction.

18.   Silica gel TLC shows one spot at R_f = 0.33 (10% ethyl acetate/ligroin).   GLC shows >95% purity; one peak at R_t = 7.04. Mass spectrum: m/e 43 (100% $COCH_3$), 97 (6.3% $M^+ - CH_2(CH_2)_4CH_3$), 125 (2.0% $M^+ - CH_2COCH_3$). IR (neat) cm^{-1}: 1700 (R'COR). ^{1}H NMR ($CDCl_3$) δ: 0.90 (t, 3 H, J = 6.5); 1.28 (m, 8 H); 1.98 (q, 2 H, J = 6.3); 2.16 (s, 3 H); 2.28 (q, 2 H, J = 6.6); 2.51 (t, 2 H, J = 7.4); 5.34-5.51 (m, 2 H). ^{13}C NMR ($CDCl_3$) δ: 14.0, 22.6, 26.9, 28.8, 29.5, 29.8, 31.7, 32.5, 43.6, 128.2, 131.5, 207.9. Anal. Calcd. for $C_{12}H_{22}O$: C, 79.06; H 12.17. Found: C, 78.79; H, 12.10.

3. Discussion

The Carroll rearrangement,[3,4] an old and well-established thermal rearrangement, involves the rearrangement of allylic esters to β-keto acids followed by decarboxylation to provide γ,δ-unsaturated methyl ketones. Even though the Carroll rearrangement is a versatile complement to the Claisen rearrangement,[5] it is not of widespread use. This may be due to: (a) the lack of a convenient, high yield method for the formation of β-keto esters and (b) the harsh conditions required to effect rearrangement.[6] Often, procedures involve direct conversion of allylic alcohols to the rearranged and decarboxylated products in one step and low yield.

The method of preparation of 5-dodecen-2-one presented here is a version of the literature procedure published earlier.[7] It offers several advantages over existing methodology: (1) The ester enolate modification of the Carroll rearrangement provides the allylic acetoacetates via a mild, fast, and high yield synthesis. This procedure represents a significant improvement over other routes.[8] (2) Dianions of the allylic acetoacetates rearrange at room temperature and the resulting β-keto acids can be readily isolated. Isolation

216

of the acetoacetic acids adds to the versatility of the synthesis of γ,δ-unsaturated methyl ketones and makes purification much more simple than the pyrolysis method. (3) Finally, the general pyrolysis procedure, although one step, leads to side products and low yields (typically 10-40%). For example, pyrolysis of (1-ethenyl)heptanyl 3-ketobutanoate[4] gives two major products, 5-dodecen-2-one and 3-hydroxy-1-nonene, whereas the method of preparation described here yields only the desired γ,δ-unsaturated methyl ketone.

The Table contains representative examples of the method of preparation described here.

1. Department of Chemistry, New York University, Washington Square, New York, NY 10003.

2. Kofron, W. G.; Baclawski, L. M. *J. Org. Chem.* **1976**, *41*, 1879.

3. Carroll, M. F. *J. Chem. Soc.* **1940**, 1266.

4. Kimel, W.; Cope, A. C. *J. Am. Chem. Soc.* **1943**, *65*, 1992.

5. Ziegler, F. E. *Acc. Chem. Res.* **1977**, *10*, 227.

6. Rearrangement is normally carried out at temperatures of 130-220°C by heating the β-keto ester neat or in a high boiling solvent (xylene, diphenyl ether), usually in situ after preparation of the β-keto ester.

7. Wilson, S. R.; Price, M. F. *J. Org. Chem.* **1984**, *49*, 722.

8. Acetoacetate formation has previously been carried out by using the following. (a) Et_3N: Kato, T.; Chiba, T. *Chem. Pharm. Bull.* **1975**, *23*, 2263; (b) NaOR: See ref. 4; (c) p-TsOH: Boese, A. B., Jr. *Ind. Eng. Chem.* **1940**, *32*, 16.

TABLE

PREPARATION OF γ,δ-UNSATURATED METHYL KETONES

SUBSTRATE	PRODUCT	YIELD	REF.
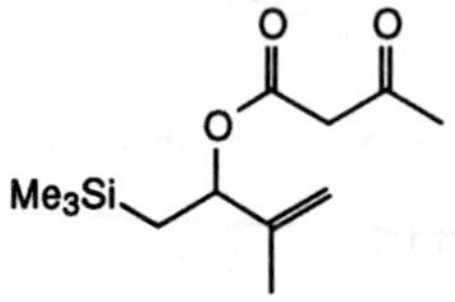	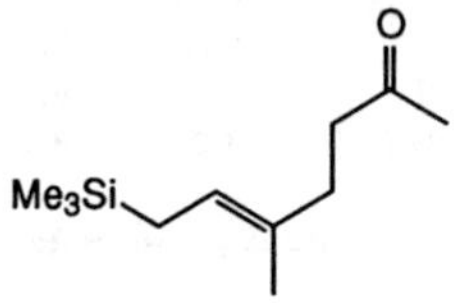	40%	7
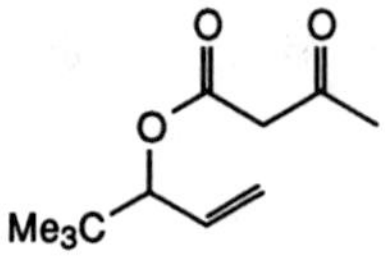	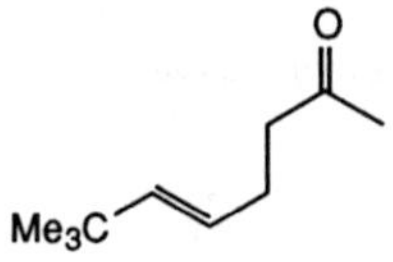	80%	7
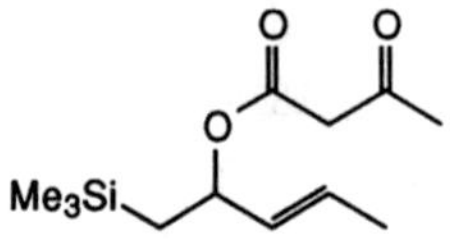	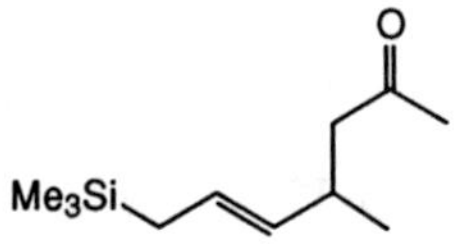	84%	7

218

Appendix

Chemical Abstracts Nomenclature (Collective Index Number); (Registry Number)

5-Dodecen-2-one: 5-Dodecen-2-one, (E)- (11); (81953-05-1).

3-Hydroxy-1-nonene: 1-Nonen-3-ol (8,9); (21964-44-3)

Diketene: 2-Oxetanone, 4-methylene- (8,9); (674-82-8)

4-Dimethylaminopyridine: Pyridine, 4-(dimethylamino)- (8); 4-Pyridinamine, N,N-dimethyl- (9); (1122-58-3)

CYCLOPENTADIENE ANNULATION VIA THE SKATTEBØL REARRANGEMENT:

(1R)-9,9-DIMETHYLTRICYCLO[6.1.1.0^{2,6}]DECA-2,5-DIENE

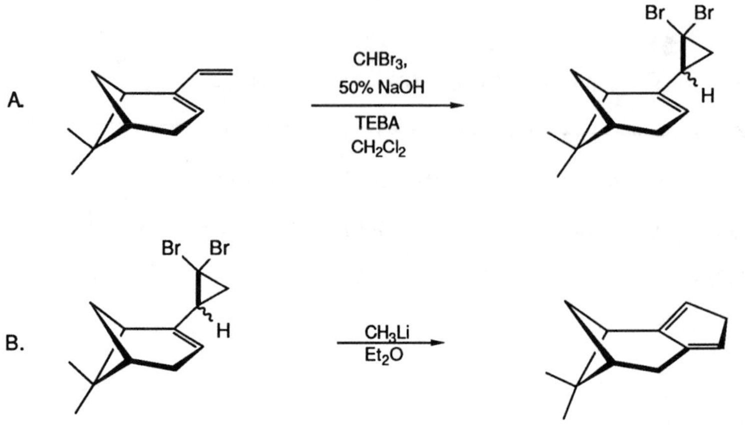

Submitted by Leo A. Paquette and Mark L. McLaughlin.[1]

Checked by Nanine Van Draanen and Clayton H. Heathcock.

1. Procedure

A. Dibromocarbene addition to (1R)-nopadiene. A 250-mL, three-necked flask is equipped with a mechanical stirrer, nitrogen inlet, and serum cap. The flask is charged with 26.2 mL (0.30 mol) of bromoform (Note 1), 29.6 g (0.20 mol) of (1R)-nopadiene (Notes 2 and 3), 1.0 g (4.4 mmol) of benzyltriethylammonium chloride (TEBA), 0.8 mL of ethanol, and 20 mL of dichloromethane (Note 4). The suspension is stirred and cooled in an ice bath while 100 mL of 50% sodium hydroxide solution is added over 10 min from a dropping funnel. The reaction mixture is stirred at room temperature for 24 hr and poured into 250 mL of water. The lower layer is separated and the aqueous phase is extracted with three 25-mL portions of dichloromethane. The combined organic layers are washed with three 100-mL portions of water, dried over magnesium sulfate, and concentrated under reduced pressure to give a brown-black oil. The oil is dissolved in an equal volume of hexane and filtered through a 2-in bed of silica with hexane (1.5 L) as eluant. The solvent is evaporated and the orange oil is distilled in an apparatus protected from light (Note 5) at 85-95°C and 0.08 mm. The yellow distillate is redistilled through a 4-in Vigreux column to give 32.0-33.6 g (50-53%) of the diastereomeric dibromocyclopropanes (Note 6).

B. *(1R)-9,9-Dimethyltricyclo[6.1.1.0^{2,6}]deca-2,5-diene.* A flame-dried, 3-L flask is equipped with a large magnetic stirring bar and serum cap and charged with 17.6 g (55.0 mmol) of the dibromide. A total of 2 L of anhydrous ether (Note 7) is transferred into the flask via cannula. The stirred solution is cooled in an ice bath and 147 mL of 1.5 M methyllithium in ether (220 mmol) is introduced via a second cannula (Note 8). The ice bath is removed and stirring is maintained for 10 hr before the solution is transferred by cannula into 1 L of ice-cold water. The ether layer is separated and the aqueous phase is extracted with two 200-mL portions of ether. The combined ethereal solutions are dried and concentrated (Note 9). The residual yellow oil is immediately diluted with an equal volume of hexane and passed through a short column of neutral alumina (Note 10). The solvent is carefully removed and the yellow oil is subjected to bulb-to-bulb distillation at 90°C and 5 mm (Note 11). The yield of colorless hydrocarbon is 6.9-7.1 g (78-80%) (Notes 12 and 13).

2. Notes

1. The submitters used a purified grade of bromoform purchased from the Fisher Chemical Company.

2. The (1R)-nopadiene is prepared from commercially available (Aldrich Chemical Company, Inc.) (1R)-(-)-nopol (Note 3) according to the following procedure.[2] A 1000-mL, three-necked flask is equipped with a mechanical stirrer, internal thermometer and nitrogen inlet. The flask is charged with 125 g (0.752 mol) of (1R)-(-)-nopol and 500 mL of pyridine. Stirring is begun and the solution is cooled to -10°C in an ice-salt bath under nitrogen. p-Toluenesulfonyl chloride (175 g, 0.918 mol) is added in one portion under an inert atmosphere via Gooch tubing (the checkers used a powder funnel for the addition). The temperature rises to 40°C for 15-20 min, but returns to 5°C where it is maintained for 2 hr. Twenty 1-mL portions of water are next introduced at such a rate that the temperature does not exceed 5°C. The reaction mixture is poured into 1 L of ether and extracted with ice-cold 5 M sulfuric acid until the aqueous layer remains acidic, then with saturated $CuSO_4$ solution until the aqueous layer remains blue. The ethereal phase is washed with two 200-mL portions each of water and 5% sodium bicarbonate solution prior to drying over magnesium sulfate and solvent evaporation. A solid residue is obtained. If this material is dark, it may be dissolved in hexane and filtered through a pad of Celite to remove the black impurity. The tosylate is recrystallized by dissolving it in 500 mL of hot hexane and cooling to -78°C. Six such recrystallizations give material with mp 51.0-51.8°C and $[\alpha]_D^{25}$ -25.6° (C_2H_5OH, c 0.03). The yield is 65-72%.

A 2000-mL, three-necked flask is equipped with a mechanical stirrer, internal thermometer, and nitrogen inlet. The flask is charged with 200 g (0.624 mol) of (1R)-nopyl tosylate and 1000 mL of dimethyl sulfoxide which has been freshly distilled from calcium hydride at 40 mm. The stirring solution is cooled briefly in a cold water bath and 69.0 g (0.615 mol) of freshly sublimed potassium tert-butoxide is added rapidly while nitrogen is flowing above the solution (the checkers used potassium tert-butoxide from a freshly-opened bottle). (The base must be the limiting reagent to offset isomerization of the product diene). The temperature rises to approximately 45°C and a brown color develops. As the reaction proceeds, the color dissipates to a light yellow. After the initial exotherm subsides, the mixture is heated at 75°C for 10 hr, cooled to room temperature, and diluted with 800 mL of hexane. The lower layer, mostly dimethyl sulfoxide, is diluted with 1 L of water and extracted with two 100-mL portions of hexane. The combined hexane layers are washed with water (5 x 200 mL), dried over magnesium sulfate, and rotary evaporated at 40 mm and 25°C to leave a yellow oil. Distillation through a 5-in Vigreux column gives 69.4-74.0 g (75-80%) of (1R)-nopadiene as a clear colorless oil, bp 78-79°C/25 mm.

3.   This compound is mislabelled and misdrawn in the 1987 and 1988 Aldrich catalogs as the S-enantiomer.

4.   These phase-transfer conditions are adapted from experimental procedures described earlier.[3,4]

5.   The dibromocyclopropane is light-sensitive when hot. Exposure to light during distillation produces colored impurities that cause autocatalytic decomposition of the product when subsequently stored in the cold.

6. Both distillations must be performed with a pot temperature below 150°C in order to avoid thermal decomposition. ^{1}H NMR indicates the product to be a 4:1 mixture of diastereomers. All available evidence denotes that both are transformed efficiently into the cyclopentadiene.

7. The ether was freshy distilled from sodium benzophenone ketyl. The checkers used anhydrous ether from a freshly-opened can.

8. The methyllithium was purchased from the Aldrich Chemical Company, Inc., and contains lithium bromide.

9. Solvent evaporation was accomplished at 40 mm and 25°C in order to counter product volatility.

10. The checkers used a 1" x 1" plug of alumina. The experience of the submitters has been that the use of silica gel at this point causes some decomposition.

11. The checkers found foaming to be a serious problem in this distillation. The problem is ameliorated by use of a 50-mL or larger distillation flask.

12. Purified diene polymerizes within 24 hr if stored neat. Its lifetime can be indefinitely prolonged by storage as a 10% by weight solution in hexane under an argon atmosphere.

13. The product exhibits $[\alpha]_D^{24}$ -21.9° (C_2H_5OH, c 1.8) and the following ^{1}H NMR spectrum at 300 MHz in $CDCl_3$ solution δ: 0.72 (s, 3 H), 1.24 (m, 1 H), 1.33 (s, 3 H), 1.60 (s, 1 H), 2.11 (m, 1 H), 2.60 (m, 1 H), 2.70 (m, 2 H), 2.99 (s, 2 H), 5.77 (s, 1 H), 5.99 (s, 1 H).

3. Discussion

Experience has shown[5,6] that cyclopentadiene annulation of 2,3-dimethylenebicyclo[2.2.2]octanes can be efficiently realized by means of the Skattebøl procedure.[7] However, the added strain in 2,3-dimethylenenorbornanes reroutes the rearrangement instead into vinylallene formation.[4] This phenomenon has been attributed to an inability on the part of the torsionally-constrained empty carbene p orbital to interact with the flanking double bond.[8] This structural inhibition is entirely alleviated by positioning the cyclopropyl carbene completely external to the norbornene ring as in the present example. The heightened conformational maneuverability of the carbenoid center is conducive to exclusive cyclopentadiene ring formation.

(1R)-9,9-Dimethyltricyclo[6.1.1.0^{2,6}]deca-2,5-diene is a chiral, optically active homolog of isodicyclopentadiene, a molecule that has been extensively studied with regard to π-facial selectivity in cycloaddition reactions.[9] The response of the title compound to similar dienophiles has been described[10] and its complexation to various transition metals reported.[10,11] The steric contributions of the gem-dimethyl substituents relegate bonding to the opposite surface of the cyclopentadiene ring.

1. Department of Chemistry, The Ohio State University, Columbus, OH 43210.

2. Cupas, C. A.; Roach, W. S. *J. Org. Chem.* **1969** *34*, 742.

3. Makosza, M.; Fedorynski, M. *Synth. Commun.* **1973**, *3*, 305.

4. Paquette, L. A.; Green, K. E.; Gleiter, R.; Schafer, W.; Gallucci, J. C. *J. Am. Chem. Soc.* **1984**, *106*, 8232.

5. Butler, D. N.; Gupta, I. *Can. J. Chem.* **1978**, *56*, 80.

6. Charumilind, P.; Paquette, L. A. *J. Am. Chem. Soc.* **1984**, *106*, 8225.

7. Skattebøl, L. *Tetrahedron* **1967**, *23*, 1107.

8. McLaughlin, M. L.; McKinney, J. A.; Paquette, L. A. *Tetrahedron Lett.* **1986**, *27*, 5595.

9. Paquette, L. A.; Gugelchuk, M.; Hsu, Y.-L. *J. Org. Chem.* **1986**, *51*, 3864 and references cited therein.

10. Paquette, L. A.; Gugelchuk, M.; McLaughlin, M. L. *J. Org. Chem.* **1987**, *52*, 4732.

11. Paquette, L. A.; McKinney, J. A.; McLaughlin, M. L.; Rheingold, A. L. *Tetrahedron Lett.* **1986**, *27*, 5599.

Appendix

Chemical Abstracts Nomenclature (Collective Index Number);
(Registry Number)

(1R)-Nopadiene: 2-Norpinene, 6,6-dimethyl-2-vinyl-, (+) (8);

Bicyclo[3.1.1]hept-2-ene, 2-ethenyl-6,6-dimethyl-, (1R)- (9); (30293-06-2)

(1R)-(-)-Nopol: Bicyclo[3.1.1]hept-2-ene-2-ethanol, 6,6-dimethyl-,

(1R)- (9); (35836-73-8)

4,13-DIAZA-18-CROWN-6

(1,4,10,13-Tetraoxa-7,16-diazacyclooctadecane)

Submitted by Vincent J. Gatto, Steven R. Miller, and George W. Gokel.[1]

Checked by P. C. Prabhakaran and James D. White.

1. Procedure

A. *1,10-Dibenzyl-4,7-dioxa-1,10-diazadecane.* A solution of benzylamine (172 g, 1.6 mol) (Note 1) and 1,2-bis(2-chloroethoxy)ethane (18.7 g, 0.1 mol) (Note 2) is stirred and heated at 120°C for 28 hr. The reaction is cooled to room temperature, sodium hydroxide pellets (8.0 g, 0.2 mol) are added, and the mixture is heated at 120°C, with stirring, for 1 hr. The reaction is cooled and excess benzylamine is removed by vacuum distillation (Note 3) using a water aspirator (Note 4). The resulting oil is dissolved in chloroform (100 mL), filtered, and washed with water (50 mL) to remove salts. The organic phase is dried (Na_2SO_4) and concentrated under reduced pressure. Bulb-to-bulb distillation using a Kugelrohr apparatus (175-177°C, 0.2 mm) gives 27.9-31.2 g

(85-95%) of 1,10-dibenzyl-4,7-dioxa-1,10-diazadecane (Note 5) as a slightly yellow oil which is sufficiently pure to be used for the preparation of N,N'-dibenzyl-4,13-diaza-18-crown-6.

B. *N,N'-Dibenzyl-4,13-diaza-18-crown-6.* In a 3-L, round-bottomed flask fitted with a mechanical stirrer and an efficient reflux condenser are placed 1,10-dibenzyl-4,7-dioxa-1,10-diazadecane (28.2 g, 86 mmol), 1,2-bis(2-iodoethoxy)ethane (39.3 g, 106 mmol) (Note 6), anhydrous sodium carbonate (45.3 g, 427 mmol), and sodium iodide (6.4 g, 43 mmol) in acetonitrile (1700 mL). The resulting solution is stirred mechanically (Note 7) and heated at reflux for 21 hr. The reaction is cooled, filtered, and concentrated under reduced pressure (Note 8). The crude solid is dissolved in a refluxing solution of acetone-dioxane (175 mL each) and allowed to crystallize in a freezer. The crystals (a mixture of sodium iodide and the sodium iodide complex of the product) are dried and taken up in 500 mL of water and 400 mL of chloroform. The phases are separated and the aqueous portion is extracted with chloroform (3 x 75 mL). The combined organic phases are dried (MgSO$_4$) and concentrated under reduced pressure. Recrystallization (hexanes, 500 mL, followed by absolute ethanol, 110 mL) affords 25.6-27.0 g (67-71%) of N,N'-dibenzyl-4,13-diaza-18-crown-6 as a white solid (mp 80-81°C); ^{1}H NMR (CDCl$_3$) δ: 2.82 (t, 8 H, NCH$_2$); 3.64 and 3.70 (t, s, s, 2 CH, OCH$_2$ and CH$_2$Ph); 7.37 (s, 10 H, Ar); IR (KBr) cm^{-1}: 2960, 2900, 2880, 1500, 1460, 1120, 1060, 1050, 750, 700 (Notes 9 and 10).

C. *4,13-Diaza-18-crown-6.* N,N'-Dibenzyl-4,13-diaza-18-crown-6 (25.0 g, 56 mmol) (Note 11), 10% Pd/C catalyst (1.0 g) and absolute ethanol (300 mL) are shaken in a Parr series 3900 hydrogenation apparatus at 60 psi hydrogen pressure and 25°C for 72 hr. The mixture is filtered through a pad of Celite and concentrated under reduced pressure. The yield of pure 4,13-diaza-18-

crown-6 after recrystallization (hexanes, 1 g/35 mL) is 13.5 g (91%). The
white solid (mp 114-115°C) possesses physical properties identical to those
previously reported:[2] ^{1}H NMR (CDCl$_3$) δ: 2.06 (bs, 2 H, NH); 2.72 (t, 8 H,
CH$_2$N); 3.54 (t, s, 16 H, CH$_2$); IR (KBr) cm^{-1}: 3330.

2. Notes

1. Benzylamine was obtained from Aldrich Chemical Company, Inc., and was
used without further purification.

2. 1,2-Bis(2-chloroethoxy)ethane was obtained from Eastman Kodak
Company, and was used without further purification.

3. The excess benzylamine, recovered from the distillation step, can be
reused if redistilled from calcium oxide.

4. If an efficient water aspirator is used (<20 mm), the benzylamine
should distill between 60 and 70°C.

5. The product has the following spectral characteristics: ^{1}H NMR
(CDCl$_3$) δ: 1.82 (s, 2 H, NH), 2.72 (t, 4 H, NCH$_2$), 3.58 (s and t, 8 H, CH$_2$),
3.72 (s, 4 H, benzyl), and 7.28 ppm (s, 10 H, Ar).

6. 1,2-Bis(2-iodoethoxy)ethane was prepared as described by Kulstad and
Malmsten.[3] 1,2-Bis(2-chloroethoxy)ethane (21.3 g, 0.114 mol) and sodium
iodide (37.0 g, 0.247 mol) in acetone (55 mL) were heated at reflux while
stirring magnetically during three days. The reaction mixture was allowed to
cool, it was filtered, and the filtrate was evaporated under reduced
pressure. The residue was dissolved in methylene chloride (200 mL), washed
with aqueous 10% sodium thiosulfate solution (2 x 100 mL), dried over
magnesium sulfate, and evaporated under reduced pressure. The residual
methylene chloride was removed by high vacuum evaporation at ambient

temperature and the resulting 1,2-bis(2-iodoethoxy)ethane (40 g, 95%) was used without further purification. The proton NMR spectrum (δ, CDCl$_3$) is as follows: 3.25 (t, 4 H); 3.68 (s, 4 H); 3.78 (t, 4 H).

7. This reaction must be stirred vigorously (120-150 rpm using a 60-mm paddle) for best results.

8. The acetonitrile in this step can be reused without any further purification.

9. Wester and Voegtle[4] reported mp 80°C.

10. When this reaction is run on twice the reported scale, the percent yield is the same.

11. The N,N'-dibenzyl-4,13-diaza-18-crown-6 must be freshly recrystallized from absolute ethanol for the hydrogenolysis to proceed at a reasonable rate.

3. Discussion

During the past two decades, a relatively few macrocyclic polyethers have played central roles in numerous research programs. Examples are 18-crown-6, dibenzo-18-crown-6, and aza-15-crown-5. 4,13-Diaza-18-crown-6 and its derivatives are compounds of considerable current interest despite the parent's high price and limited availability. 4,13-Diaza-18-crown-6 is a key compound in the study of two-armed macrocycles since it may readily be alkylated or acylated to afford a variety of symmetrical, N,N'-disubstituted derivatives.

4,13-Diaza-18-crown-6 has been prepared in a variety of ways.[2-3,5-9] Lehn first reported its preparation by reaction of 1,2-bis(2-aminoethoxy)ethane with triglycolic acid dichloride, followed by lithium aluminum hydride or diborane reduction of the resulting bislactam.[2] Kulstad and Malmsten have condensed 1,2-bis(2-aminoethoxy)ethane with 1,2-bis(2-iodoethoxy)ethane to give 4,13-diaza-18-crown-6.[3] Recently, we have reported that a single-step reaction of benzylamine with 1,2-bis(2-iodoethoxy)ethane, followed by hydrogenation of the resulting N,N'-dibenzyl-protected crown, gives 4,13-diaza-18-crown-6.[5] The latter, single-step cyclization reaction is more direct than the present procedure, but the yield is substantially lower and the manipulations are less convenient.

The method described here offers three advantages over the previously published procedures.[2-3,5-9] First, the cyclization reaction does not require the use of high dilution conditions in order to obtain satisfactory yields of product. This is a substantial improvement over the procedure of Lehn[2] which requires large volumes of dry solvents and slow addition rates. Second, purification of all the intermediates is straightforward, involving either vacuum distillation using a Kugelrohr apparatus or recrystallization. This is an important advantage when the sequence is scaled up because it allows the preparation of large sample sizes in relatively short periods of time. We have prepared as much as 40 g of 4,13-diaza-18-crown-6 in less than 1 week. Third, the benzyl protecting groups are easily removed by hydrogenolysis over H_2/Pd-C in ethanol. Previous preparations of 4,13-diaza-18-crown-6 are more difficult because they involve the hydrolysis or reduction of N-tosyl protected nitrogens.[5-8] We should also note that our own previously published,[5] single-step preparation for N,N-disubstituted-4,13-diaza-18-crown-6 derivatives is more convenient than the present preparation because it

231

involves a single step reaction, but the yields are always inferior to those
obtained using the present, multi-step approach.

1. Department of Chemistry, University of Miami, Coral Gables, FL 33124.

2. Dietrich, B.; Lehn, J. M.; Sauvage, J. P. *Tetrahedron Lett.* **1969**, 2885.

3. Kulstad, S.; Malmsten, L. A.; *Acta. Chem. Scand., Ser. B* **1979**, *B33*, 469.

4. Wester, N., Voegtle, F. *J. Chem. Res. (S)* **1978**, 400.

5. Gatto, V. J.; Gokel, G. W. *J. Am. Chem. Soc.* **1984**, *106*, 8240.

6. Buhleier, E.; Rasshofer, W.; Wehner, W.; Luppertz, F.; Voegtle, F. *Justus
   Liebigs Ann. Chem.* **1977**, 1344.

7. Desreux, J. F.; Renard, A.; Duyckaerts, G. *J. Inorg. Nucl. Chem.* **1977**,
   39, 1587.

9. Bogatsky, A. V.; Lukyanenko, N. G.; Basok, S. S.; Ostrovskaya, L. K.
   Synthesis **1984**, 138.

9. Richman, J. E.; Atkins, T. J. *J. Am. Chem. Soc.* **1974**, *96*, 2268.

Appendix

Chemical Abstracts Nomenclature (Collective Index Number);

(Registry Number)

4,13-Diaza-18-crown-6: 1,4,10,13-Tetraoxa-7,16-diazacyclooctadecane (8,9);
(23978-55-4)

1,10-Dibenzyl-4,7-dioxa-1,10-diazadecane: Benzenemethanamine,
N,N'-[1,2-ethanediylbis(oxy-2,1-ethanediyl)]bis- (10); (66582-26-1)

1,2-Bis(2-chloroethoxy)ethane: Ethane, 1,2-bis(2-chloroethoxy)- (8,9);
(112-26-5)

N,N'-Dibenzyl-4,13-diaza-18-crown-6: 1,4,10,13-Tetraoxa-7,16-
diazacyclooctadecane, 7,16-bis(phenylmethyl)- (10); (69703-25-9)

1,2-Bis(2-iodoethoxy)ethane: Ethane, 1,2-bis(2-iodoethoxy)- (9); (36839-55-1)

p-tert-BUTYLCALIX[4]ARENE

Submitted by C. D. Gutsche and M. Iqbal.[1]

Checked by Anthony T. Watson and Clayton H. Heathcock.

1. Procedure

A. Preparation of "precursor". A solution prepared from 100 g (0.666 mol) of p-tert-butylphenol, 62 mL of 37% formaldehyde solution (0.83 mol of HCHO), and 1.2 g (0.03 mol) of sodium hydroxide (corresponding to 0.045 equiv with respect to the phenol) (Note 1) is placed in a 3-L, three-necked flask equipped with a mechanical stirrer. The contents of the open flask are heated by means of a heating mantle (Note 2) for ca. 2 hr at 110-120°C. The reaction mixture, which is clear at the beginning, becomes viscous and turns yellow, eventually changing to a thick light yellow mass as the water evaporates (Note 3). During this period there is considerable frothing, and the reaction mixture may fill most of the flask before shrinking back to the original volume. The reaction vessel is removed from the heating mantle, allowed to cool to room temperature, and 800-1000 mL of warm diphenyl ether is added to the flask to dissolve the residue. This process, which is facilitated by stirring, generally requires about 1 hr.

B. *Pyrolysis of the precursor*. The 3-L, three-necked flask is fitted with a nitrogen inlet. The contents of the flask are heated with a heating mantle while a stream of nitrogen is blown rapidly over the reaction mixture to facilitate the removal of the water that is evolved. During this period the color of the solution changes from yellow to light brown. When the evolution of water subsides and a solid starts to form (prior to attaining the reflux temperature) (Note 4) the flask is fitted with a condenser, and the contents of the flask are refluxed for 1.5-2 hr. During this phase of the reaction the solid dissolves, and a clear dark brown solution is formed. The reaction mixture is cooled to room temperature, and the product is precipitated by treatment with 1.5 L of ethyl acetate. The resulting mixture is stirred for 15-30 min and allowed to stand for at least 30 min (Note 5). Filtration yields material which is washed twice with 100-mL portions of ethyl acetate, once with 200 mL of acetic acid, and twice with 100-mL portions of water to yield ca. 66 g (61%) of crude product (Note 6). The beige-colored crude product is dissolved in ca. 1600-1800 mL of boiling toluene which is concentrated to ca. 700-900 mL. Upon cooling, 61 g (49%) of product is obtained as glistening white rhombic crystals, mp 342-344°C (Notes 7 and 8).

2. Notes

1. p-tert-Butylphenol from Aldrich Chemical Company, Inc., mp 98-101°C, and 37% formaldehyde solution from Fisher Chemical Company, Certified ACS grade, were used.

2. Care should be taken not to allow the heating mantle to get so hot as to char the solid material on the walls of the flask. Submitters have used an oil bath.

3. Stirring accelerates the rate of water removal, but is not necessary.

4. In some cases a solid does not form, and the solution remains clear throughout.

5. It may be convenient to transfer the contents of the three-necked flask to an Erlenmeyer flask prior to the addition of ethyl acetate.

6. The crude material is usually pure enough to be used in subsequent reactions without recrystallization.

7. The product is a 1:1 complex of p-tert-butylcalix[4]arene and toluene, from which the toluene can be removed by drying under high vacuum (>1 mm) and high temperature (>140°C) for an extended period of time (48 hr).

8. The melting point is measured in an evacuated melting point tube.

3. Discussion

A prototype of this procedure was first published in 1941 by Zinke and Ziegler,[2] although the unambiguous identity of the product was not established until some years later.[3] A few other p-alkylcalix[4]arenes have been prepared by this procedure, but for practical purposes it appears to be restricted to p-alkylphenols in which the p-alkyl group is highly branched at the position adjacent to the phenyl ring. Thus, p-tert-pentylcalix[4]arene and p-(1,1,3,3-tetramethylbutyl)calix[4]arene[4] are among the few other phenols that yield a tractable product in reasonable yield.

1. Department of Chemistry, Washington University, St. Louis, MO 63130.

2. Zinke, A.; Ziegler, E. *Ber.* **1941**, *74B*, 1729; Zinke, A.; Ziegler, E.; Martinowitz, E.; Pichelmayer, H.; Tomio, M.; Wittmann-Zinke, H.; Zwanziger, S. *Ber.* **1944**, *77B*, 264.

3. Gutsche, C. D.; Dhawan, B.; No, K. H.; Muthukrishnan, R. *J. Am. Chem. Soc.* **1981**, *103*, 3782.

4. Foina, D.; Pochini, A.; Ungaro, R. Andreetti, G. D. *Makromol. Chem., Rapid. Commun.* **1983**, *4*, 71.

Appendix
Chemical Abstracts Nomenclature (Collective Index Number);
(Registry Number)

p-tert-Butylcalix[4]arene: Pentacyclo[19.3.1.1^{3,7},1^{9,13},1^{15,19}]octacosa-1(25),3,5,7(28),9,11,13(27),15,17,19(26),21,23-dodecaene-25,26,27,28-tetraol, 5,11,17,23-tetrakis(1,1-dimethylethyl)- (10); (60705-62-6)

p-tert-Butylphenol: Phenol, p-tert-butyl- (8); Phenol, 4-(1,1-dimethylethyl)- (9); (98-54-4)

p-tert-BUTYLCALIX[6]ARENE

Submitted by C. D. Gutsche, B. Dhawan, M. Leonis, and D. Stewart.[1]

Checked by Roger J. Butlin and James D. White.

1. Procedure

A 2-L, three-necked, round-bottomed flask equipped with a nitrogen inlet, mechanical stirrer, and a Dean-Stark trap and condenser (Note 1), is placed in a Glas-Col heating mantle. To the flask are added 100 g (0.665 mol) of p-tert-butylphenol, 135 mL of 37% formalin solution (1.8 mol of HCHO), and 15 g (0.227 mol) of potassium hydroxide pellets (corresponding to 0.34 equiv of the phenol) (Notes 2 and 3). Heating and stirring are begun, and after 15 min nitrogen is blown across the reaction mixture at a brisk rate and out through the condenser on top of the Dean-Stark trap (Note 4); the reaction mixture is heated and stirred for 2 hr (Note 5). As the reaction progresses, the originally clear solution turns bright lemon yellow and as water is removed, the reaction mixture eventually changes to a thick, golden yellow mass of taffy-like consistency (Note 6). During this period some frothing occurs, and the reaction mixture expands somewhat before shrinking to the original volume. Xylene (1 L) is now added to the flask to dissolve the semi-solid mass and give a yellow solution which is brought quickly to reflux by increasing the temperature of the heating mantle (Note 7). After 30 min a precipitate begins to form, and the color of the reaction mixture changes from yellow to orange. Refluxing is continued for 3 hr, the heating mantle is removed, and the mixture is allowed to cool to room temperature. The mixture is filtered, and the precipitate is washed with xylene and dried on a Büchner funnel to yield 105-110 g of crude, almost colorless product. This material is powdered, placed in an Erlenmeyer flask, dissolved in 2.5 L of choloroform (not completely soluble), and treated with 800 mL of 1 N hydrochloric acid. After 10-15 min the stirred solution turns yellow to light orange; stirring is continued an additional 10 min, and the mixture is transferred to a separatory funnel. The chloroform layer is drawn off, the aqueous layer is extracted with an additional 250 mL of chloroform, and the combined chloroform extracts are washed once with water and dried over magnesium sulfate. Magnesium sulfate is removed by filtration, the chloroform solution is concentrated to ca. 1 L by boiling, and 1 L of hot acetone is added to the boiling chloroform solution. The mixture is allowed to cool and is then filtered to give 90-95 g (83-88%) of product as a white powder: mp 372-374°C (Notes 8, 9, and 10).

2. Notes

1. It is not essential that a Dean-Stark trap be used. However, it provides a convenient way for monitoring this stage of the reaction and for collecting the aqueous formaldehyde mixture that is evolved.

239

2. p-tert-Butylphenol from Aldrich Chemical Company, Inc., mp 98-101°C, and 37% formaldehyde solution from Fisher Chemical Company, Certified ACS grade, were used. The potassium hydroxide (KOH) pellets are ca. 85% KOH and 15% water. When sodium hydroxide or lithium hydroxide is used instead of potassium hydroxide the yield of product is considerably lower.

3. Sometimes the reaction between the phenol and formaldehyde proceeds with sufficient speed after mixing that considerable warming takes place. Whether or not spontaneous warming occurs, the ultimate outcome of the reaction is the same.

4. Typically, the rate of nitrogen flow is ca. 200 bubbles/min, as measured by a bubbler attached to the condenser outlet. Although a rapid flow of nitrogen is not essential, it facilitates the removal of water and formaldehyde from the reaction mixture; this should amount to ca. 85 mL in the lower layer of the Dean-Stark trap.

5. Stirring is not absolutely necessary, but it accelerates the rate of removal of water. In unstirred reaction mixtures the frothing contents may fill the entire flask. Stirring reduces the amount of frothing.

6. As the reaction mixture becomes more viscous it is necessary to increase the torque on the stirring motor to keep the taffy-like mass in motion while heating is continued. It is important to remove as much of the water as possible in this first stage of the reaction by continuing the stirring as long as possible (see Note 7). When stirring becomes very difficult or when the yellow mass no longer sticks to the sides of the flask, xylene is added. Care should be taken not to scorch the reaction mixture at this point. As a further aid to rapid removal of water it is recommended that the upper portion of the reaction flask as well as the Dean-Stark trap be covered with insulating material such as cotton or glass wool.

7.   Incomplete removal of water and/or too slow removal of water can result in the formation of a significant amount of cyclic octamer, which is difficult to separate from the cyclic hexamer. It is important, therefore, to bring the reaction mixture to reflux (10-15 min) and to remove all the water. In addition to the ca. 85 mL of H_2O/HCHO removed in the first phase of the reaction, an additional ca. 10 mL (as a cloudy, lower layer in the Dean-Stark trap) is removed in the second phase.

8.   HPLC analysis of this product usually shows it to contain less than 1% of other calixarenes. If these are present in greater amounts (usually the cyclic tetramer, cyclic heptamer or cyclic octamer), their content can be reduced to less than 1% by triturating the product with hot acetone. The purity of the product can be qualitatively checked with TLC using a petroleum ether (30-60°C)/methylene chloride mixture (1:1) as eluant. The R_f of the cyclic hexamer is ca. 0.65, and the appearance of any spots with R_f greater than 0.1 indicates the presence of other calixarenes.

9.   The melting point is measured in an evacuated and sealed melting point tube.

10.   ^{1}H NMR spectrum (400 MHz, $CDCl_3$) δ: 1.29 (s, 54 H), 3.90 (s, 12 H), 7.16 (s, 12 H), 10.42 (s, 6 H).

3. Discussion

p-tert-Butylcalix[6]arene was first prepared by Gutsche, et al.[2] by the reaction of p-tert-butylphenol and paraformaldehyde in the presence of rubidium hydroxide (RbOH). A few other p-substituted calix[6]arenes have been prepared, including p-phenylcalix[6]arene[3] and p-isopropylcalix[6]arene.[4]

1. Department of Chemistry, Washington University, St. Louis, MO 63130

2. Gutsche, C. D.; Dhawan, B.; No, K. H.; Muthukrishnan, R. *J. Am. Chem. Soc.* **1981**, *103*, 3782.

3. Gutsche, C. D.; Pagoria, P. F.; *J. Org. Chem.* **1985**, *50*, 5795.

4. Dhawan, B.; Chen, S. I.; Gutsche, C. D. *Macromol. Chem.* **1987**, *188*, 921.

Appendix
Chemical Abstracts Nomenclature (Collective Index Number);
(Registry Number)

p-tert-Butylcalix[6]arene: Heptacyclo[31.3.1.1^{3,7}.1^{9,13}.1^{15,19}.-1^{21,25}.1$^{27.31}$]octetraconta-1(37),3,5,7(42),9,11,13(41),15,17,-19(40),21,23,25(39),27,29,31(38),33,35-octadecaene-37,38,39,40,41,42-hexol,5,11,17,23,29,35-hexakis (1,1-dimethylethyl)- (10); [78092-53-2]

p-tert-Butylphenol: Phenol, p-tert-butyl- (8); Phenol, 4-(1,1-dimethylethyl)-(9); (98-54-4)

Formaldehyde (8,9); (50-00-0)

p-tert-BUTYLCALIX[8]ARENE

Submitted by J. H. Munch[1a] and C. D. Gutsche.[1b]
Checked by Daniel T. Daly and James D. White.

1. Procedure

A slurry prepared from 100 g (0.67 mol) of p-tert-butylphenol, 35 g (ca. 1.1 mol) of paraformaldehyde (Note 1), and 2.0 mL (0.02 mol) of 10 N sodium hydroxide (Note 2) in 600 mL of xylene is placed in a 2-L, round-bottomed, three-necked flask fitted with a Dean-Stark water collector and a mechanical stirrer. The air in the flask is replaced with nitrogen, and the stirred contents of the flask are heated to reflux by means of a heating mantle. After 30 min a homogeneous solution is obtained, and after 1 hr a white precipitate begins to form. The reaction mixture is refluxed for 4 hr, the heating mantle is removed, the mixture is allowed to cool to room temperature, and the precipitate is removed by filtration. The crude product is washed, in succession, with 400-mL portions of toluene, ether, acetone, and water and is

243

then dried under reduced pressure. It is dissolved in ca. 1600 mL of boiling chloroform, the chloroform is concentrated to ca. 1200 mL, the solution is cooled to room temperature, and the precipitate is collected by filtration to yield 67-70 g (62-65%) of a colorless powder, dec 418-420°C (Notes 3 and 4).

2. Notes

1. p-tert-Butylphenol from Aldrich Chemical Company, Inc., mp 98-101°C, and paraformaldehyde from Fisher Chemical Company, Certified ACS grade, were used.

2. Other bases, including KOH, RbOH, and CsOH, also work with approximately the same results, but LiOH is considerably inferior.

3. The (solvated) product can be obtained in crystalline form but, upon standing in air for a few minutes, the colorless, glistening needles change to a white powder as the result of loss of solvent of crystallization. Considerable variation in the melting point of this material is noted. The product generally melts above 400°C, but sometimes the melting point falls to ca. 395°C or even lower. Undoubtedly, this is due to impurities which may be metal ions and/or other cyclic oligomers that are incompletely removed in the recrystallization (see Note 4).

4. Evaporation of the xylene filtrate and trituration of the residue with methylene chloride yields ca. 11 g of solid which consists mainly of cyclic hexamer and cyclic tetramer. The methylene chloride contains inter alia, cyclic pentamer, cyclic heptamer, and bishomo compound, and these can be obtained as pure samples in low yield by fractional crystallization procedures. The compositon of the reaction mixtures can be qualitatively established by TLC by means of the following R_f values in 9:1 petroleum

ether/acetone and 1:1 petroleum ether/methylene-chloride, respectively: (a) cyclic octamer - 0.54, 0.85, (b) cyclic heptamer - 0.40, 0.78, (c) cyclic hexamer - 0.54, 0.76, (d) cyclic pentamer - 0.74, 0.68, (e) cyclic tetramer - 0.63, 0.66, and (f) bishomooxa compound - 0.66, 0.56.

3. Discussion

This method for preparing p-tert-butylcalix[8]arene was first described in the patent literature[2] by chemists of the Petrolite Corporation, St. Louis, MO and, therefore, is often referred to as the "Petrolite Procedure". It was introduced into journal literature by Gutsche, et al.[3] Although the procedure is restricted to phenols substituted in the p-position with electronically neutral groups, it is more general in its application than the accompanying procedures for the calix[4]arenes and calix[6]arenes, and has been used with p-isopropylphenol,[4] p-tert-pentylphenol,[4] p-(1,1,3,3-tetramethylbutyl)phenol,[5] and p-phenylphenol.[6]

1. (a) Petrolite Corporation, St. Louis, MO 63119; (b) Department of Chemistry, Washington University, St. Louis, MO 63130.

2. Buriks, R. S.; Fauke, A. R.; Munch, J. H. U.S. Patent 4 259 464, 1981; *Chem. Abstr.* **1981**, *94*, P209, 722x.

3. Gutsche, C. D.; Dhawan, B.; No, K. H.; Muthukrishnan, R. *J. Am. Chem. Soc.* **1981**, *103*, 3782.

4. Dhawan, B.; Chen, S. I.; Gutsche, C. D. *Makromol. Chem.* **1987**, *188*, 921.

5. Cornforth, J. W.; D'Arcy Hart, P.; Nicholls, G. A.; Rees, R. J. W.; Stock, J. A. *Brit. J. Pharmacol.* **1955**, *10*, 73.

6. Gutsche, C. D.; Pagoria, P. F. *J. Org. Chem.* **1985**, 50, 5795.

Chemical Abstracts Nomenclature (Collective Index Number);

(Registry Number)

p-tert-Butylcalix[8]arene: Nonacyclo[43.3.1.1^{3,7}.1^{9,13}.1^{15,19}.1^{21,25}.1^{27,31}_.1^{33,37}.1^{39,43}]hexapentaconta-1(49),3,5,7(56),9,11,13(55),15,17, 19(54),21,23,25(53),27,29,31(52),33,35,37(51),39,41,43(50),45,47-tetracosaene-49,50,51,52,53,54,55,56-octol, 5,11,17,23,29,35,41,47-octakis(1,1-dimethyl ethyl)- (10); (68971-82-4)

p-tert-Butylphenol: Phenol, p-tert-butyl- (8); Phenol, 4-(1,1-dimethylethyl)- (9); (98-54-4)

Paraformaldehyde (9); (30525-89-4)

Unchecked Procedures

Accepted for checking during the period August 1, 1988

through May 1, 1989. An asterisk (*) indicates that

the procedure has been subsequently checked.

In accordance with a policy adopted by the Board of Editors, beginning with Volume 50 and further modified subsequently, procedures received by the Secretary and subsequently accepted for checking will be made available upon request to the Secretary, if the request is accompanied by a stamped, self-addressed envelope. (Most manuscripts require 54¢ postage).

Address requests to:

Professor Jeremiah P. Freeman
Organic Syntheses, Inc.
Department of Chemistry
University of Notre Dame
Notre Dame, Indiana 46556

It should be emphasized that the procedures which are being made available are unedited and have been reproduced just as they were first received from the submitters. There is no assurance that the procedures listed here will ultimately check in the form available, and some of them may be rejected for publication in *Organic Syntheses* during or after the checking process. For this reason, *Organic Syntheses* can provide no assurance whatsoever that the procedures will work as described and offers no comment as to what safety hazards may be involved. Consequently, more than usual caution should be employed in following the directions in the procedures.

Organic Syntheses welcomes, on a strictly voluntary basis, comments from persons who attempt to carry out the procedures. For this purpose, a Checker's Report form will be mailed out with each unchecked procedure ordered. Procedures which have been checked by or under the supervision of a member of the Board of Editors will continue to be published in the volumes of *Organic Syntheses*, as in the past. It is anticipated that many of the procedures in the list will be published (often in revised form) in *Organic Syntheses* in future volumes.

2460R* Enantioselective Saponification with Pig Liver Esterase (PLE)
        (1S,2S,3R)-3-Hydroxy-2-nitrocyclohexyl Acetate
        M. Eberle, M. Missbach, and D. Seebach,
        Laboratorium für Organische Chemie, Eidgenössische Technische
        Hochschule, Universitätstr. 16, CH - 8092 Zürich, Switzerland

2500*   Hydromagnesiation Reaction of Propargylic Alcohols
        F. Sato and Y. Kobayashi, Department of Chemical Engineering,
        Tokyo Institute of Technology, Meguro, Tokyo 152, Japan

2501R   A General Synthetic Method for the Oxidation of Primary Alcohols to
        Aldehydes: S(+)-2-Methylbutanal
        P. L. Anelli, F. Montanari, and S. Quici, Centro CNR and
        Dipartimento di Chimica Organica e Industriale
        dell'Universita, Via Golgi 19, I-20133 Milano, Italy

2502*   1,3,4,6-Tetra-O-acetyl-2-deoxy-α-D-glucopyranose
        B. Giese and K. S. Gröninger, Institut für Organische Chemie
        und Biochemie der Technischen Hochschule, Petersenstrasse 22,
        D-6100 Darmstadt, Federal Republic of Germany

2503R* Asymmetric Synthesis of 4,4-Dialkylcyclohexenones from Chiral
        Bicyclic Lactams. (S)-4-Ethyl-4-allyl-2-cyclohexenone
        A. I. Meyers and D. Berney, Department of Chemistry,
        Colorado State University, Fort Collins, CO 80523

2504*   9-n-Butyl-1,2,3,4,5,6,7,8-Octahydroacridin-4-ol
        T. W. Bell, Y.-M. Cho, A. Firestone, K. Healy, J. Liu, R. Ludwig and
        S. D. Rothenberger, Department of Chemistry, State University of
        New York at Stony Brook, Stony Brook, NY 11794-3400

2505*   Methyl (Z)-3-(Phenylsulfonyl)-prop-2-enoate
        G. C. Hirst and P. J. Parsons, Department of Chemistry
        University of California, Irvine, Irvine, CA 92717

2506*   Lipase Catalyzed Kinetic Resolution of Alcohols via Chloroacetate
        Esters: (-)-(1R,2S)-trans-2-Phenylcyclohexanol and (+)-(1S,2R)-
        trans-2-Phenylcyclohexanol
        A. Schwartz, P. Madan, J. K. Whitesell, and R. M. Lawrence,
        Department of Chemistry, University of Texas, Austin,
        Austin, TX 78712

2507*   A Useful Fluorinating Agent, N-Fluoropyridinium Triflate, and α-
        Fluorination of a Ketone
        T. Umemoto, K. Tomita, and K. Kawada, Sagami Chemical Research
        Center, Nishi-Ohnuma 4-4-1, Sagamihara, Kanagawa 229, Japan

2510   Intramolecular Oxidative Coupling of a Bisenolate: 4-
        Methyltricyclo[$2.2.2.0^{3,5}$]octane-2,6-dione
        M.-A. Poupart, G. Lassalle, and L. A. Paquette, Department of
        Chemistry, The Ohio State University, Columbus, OH 43210

2511*   Mixed Higher Order Cyanocuprate-Induced Epoxide Openings:
          1-Benzyloxy-4-penten-1-ol
          B. H. Lipshutz, R. Moretti, and R. Crow, Department of Chemistry
          University of California, Santa Barbara, CA 93106

2512*   A General Method for the Preparation of 9-, 10-, and 11-Membered
          Unsaturated Macrolides: Synthesis of 8-Propionyl-E-5-nonenolide
          K. S. Webb, E. Asirvatham, and G. H. Posner, Department of Chemistry
          The Johns Hopkins University, Baltimore, MD 21218

2513*   The Conversion of Esters to Allylsilanes: Trimethyl(2-methylene-4-
          phenyl-3-butenyl)silane
          W. H. Bunnelle and B. A. Narayanan, Department of Chemistry
          University of Missouri, Columbia, MO 65211

2514*   Alkynyl(phenyl)iodonium Tosylates: Preparation and Stereospecific
          Coupling with Vinylcopper Reagents. Formation of Conjugated Enynes
          P. J. Stang and T. Kitamura, Department of Chemistry
          University of Utah, Salt Lake City, UT 84112

2515   Diazoketone Cyclization onto a Benzene Ring: 3,4-Dihydro-1(2H)-
          Azulenone
          L. T. Scott and C. A. Sumpter, Department of Chemistry
          University of Nevada-Reno, Reno, NV 89557-0020

2516   Oxidation of Secondary Amines to Nitrones: 6-Methyl-2,3,4,5-
          Tetrahydropyridine N-Oxide
          S.-I. Murahashi, T. Shiota, and Y. Imada, Department of Chemistry
          Faculty of Engineering Science, Osaka University
          Machikaneyama, Toyonaka, Osaka, Japan 560

2518   Preparation of (E,Z)-1-Methoxy-2-methyl-3-(trimethylsilyloxy)-1,3-
          pentadiene
          D. C. Myles and M. H. Bigham, Department of Chemistry
          Yale University, New Haven, CT 06511

2519*   Synthesis of β-Ketoesters by C-Acylation of Preformed Enolates with
          Methyl Cyanoformate: Preparation of (1α,4aβ,8aα)Methyl 2-Oxo-
          decahydro-1-naphthoate
          S. R. Crabtree, L. N. Mander, and S. P. Sethi, Research School of
          Chemistry, Australian National University, GPO Box 4
          Canberra, A.C.T. 2601, Australia

2520   9-Thiabicyclo[3.3.1]nonane-2,6-dione
          R. Bishop, School of Chemistry, The University of New South Wales
          P.O. Box 1, Kensington, New South Wales, Australia 2033

2521   Synthesis of N-Protected α-Amino Acids from N-(Benzyloxycarbonyl)-L-
          Serine via Its β-Lactone: N^{α}-(Benzyloxycarbonyl)-β-(Pyrazol-1-yl)-
          L-Alanine
          S. V. Pansare, G. Huyer, L. D. Arnold, and J. C. Vederas, Department
          of Chemistry, University of Alberta, Edmonton, Alberta,
          Canada T6G 2G2

2523   2-(2,2-Dimethoxyethyl)cyclopentanone
        E. Baciocchi and R. Ruzziconi, Dipartimento di Chimica,
        Universita di Perugia, 06100 Perugia, Italy

2524   Preparation of Enantiomerically Pure (2R)- and (2S)-Ethyl-2-
        Fluorohexanoate by Enzyme-Catalyzed Kinetic Resolution
        P. Kalaritis and R. W. Regenye, Abbott Laboratories,
        Bldg. R1B, Rm. 2047, N. Chicago, IL 60064

2526   2-Substituted Pyrroles from N-tert-Butoxycarbonyl-2-bromopyrrole:
        N-tert-Butoxy-2-trimethylsilylpyrrole
        W. Chen, E. K. Stephenson, and M. P. Cava, Department of Chemistry,
        University of Alabama, Tuscalousa, AL 35487

2528   (E)-1-Benzyl-3-(1-Iodoethylidene)piperidine: Nucleophile-Promoted
        Alkyne-Iminium Ion Cyclizations
        H. Arnold, L. E. Overman, M. J. Sharp, and M. C. Witschel,
        Department of Chemistry, University of California, Irvine,
        Irvine, CA 92717

2530   2-Cyano-6-phenyloxazolopiperidine
        M. Bonin, D. S. Grierson, J. Royer, and H.-P. Husson,
        Institut de Chimie des Substances, Naturelles du C.N.R.S.,
        91198 Gif-sur-Yvette Cedex, France

2531   1,1-Dimethylethyl (S)- or (R)-4-Formyl-2,2-dimethyl-3-
        oxazolidinecarboxylate: A Useful Serinal Derivative
        P. Garner and J. M. Park, Department of Chemistry,
        Case Western Reserve University, Cleveland, OH 44106

CUMULATIVE AUTHOR INDEX
FOR VOLUMES 65, 66, 67, AND 68

This index comprises the names of contributors to Volumes **65, 66, 67**, and **68** only. For authors to previous volumes, see either indices in Collective Volumes I through VII or the single volume entitled *Organic Syntheses, Collective Volumes, I, II, III, IV, V, Cumulative Indices,* edited by R. L. Shriner and R. H. Shriner

Taylor, S. J., **68**, 8
Thomas, A. J., **66**, 194
Thompson, E. A., **66**, 132
Threlkel, R. S., **65**, 42
Truesdale, L., **67**, 13
Tasi, Y.-M., **66**, 1, 8, 14
Tsuboi, S., **66**, 22

Urabe, H., **66**

Vaid, R. K., **68**, 175
Van Deusen, S., **68**, 92
Varghese, V., **67**, 141
Villieras, J., **66**, 220
Vishwakarma, L. C., **66**, 203
Vogel, D. E., **66**, 29

Walshe, N. D. A., **65**, 1
Watanabe, S., **67**, 44
Waykole, L., **67**, 149
Werley, Jr., R. T., **68**, 14
Wester, R. T., **65**, 108
Wilson, S. R., **68**, 210
Woodside, A. B., **65**, 211
Woodward, F. E., **65**, 1

Yamada, T., **67**, 48
Yamagata, T., **67**, 33
Yamamoto, H., **66**, 185; **67**, 76, 176
Yasuda, M., **66**, 142
Yoshida, Z., **67**, 98

Zhao, S. H., **68**, 49
Ziegler, F. E., **65**, 108

CUMULATIVE SUBJECT INDEX
FOR VOLUMES 65, 66, 67, AND 68

This index comprises subject matter for Volumes **65, 66, 67**, and **68**. For subjects in previous volumes, see either the indices in Collective Volumes I through VII or the single volume entitled *Organic Syntheses, Collective Volumes I, II, III, IV, V, Cumulative Indices*, edited by R. L. Shriner and R. H. Shriner.

The index lists the names of compounds in two forms. The first is the name used commonly in procedures. The second is the systematic name according to **Chemical Abstracts** nomenclature, accompanied by its registry number in parentheses. While the systematic name is indexed separately, it also accompanies the common name. Also included are general terms for classes of compounds, types of reactions, special apparatus, and unfamiliar methods.

Most chemicals used in the procedure will appear in the index as written in the text. There generally will be entries for all starting materials, reagents, intermediates, important by-products, and final products. Entries in capital letters indicate compounds, reactions, or methods appearing in the title of the preparation.

Acetamide, N,N-dimethyl-, **67**, 98

Acetamide, N-hydroxy-N-phenyl-, **67**, 187

Acetic acid, chloro-, 5-methyl-2-(1-methyl-1-phenylethyl)cyclohexyl ester,
[1R-(1α,2β,5α)]-, **65**, 203

Acetic acid, (diethoxyphosphinyl)-, ethyl ester; (867-13-0), **66**, 224

Acetic acid ethenyl ester, **65**, 135

Acetic acid, (hexahydro-2H-azepin-2-ylidene)-, ethyl ester, (Z)-, **67**, 170

Acetic acid, hydrazinoimino-, ethyl ester; (53085-26-0), **66**, 149

Acetic acid, lithium salt, dihydrate, **67**, 105

Acetic acid, palladium(2+) salt, **67**, 105

Acetic acid, trifluoro-, anhydride, **65**, 12

Acetic acid vinyl ester, **65**, 135

Acetoacetylation, with diketene, **68**, 210

Acetohydroxamic acid, N-phenyl-, **67**, 187

ACETONE TRIMETHYLSILYL ENOL ETHER: SILANE, (ISOPROPENYLOXY)-
TRIMETHYL-; SILANE, TRIMETHYL[(1-METHYLETHENYL)OXY]-;
(1833-53-0), **65**, 1

Acetonitrile, purification, **66**, 101

Acetophenone; Ethanone, 1-phenyl-; (98-86-2), **65**, 6, 119

Acetophenone silyl enol ether: Silane, trimethyl[(1-phenylvinyl)oxy]-;
Silane, trimethyl[(1-phenylethenyl)oxy]-; (13735-81-4), **65**, 12

4-ACETOXYAZETIDIN-2-ONE: 2-AZETIDINONE, 4-HYDROXY-ACETATE (ESTER):
2-AZETIDINONE, 4-(ACETYLOXY)-; (28562-53-0), **65**, 135

1-ACETOXY-4-BENZYLAMINO-2-BUTENE: 2-BUTEN-1-OL, 4-CHLORO-,
ACETATE (E)-; (34414-28-3), **67**, 105

4-Acetoxy-3-chloro-1-butene: 3-Buten-1-ol, 2-chloro-, acetate; (96039-67-7), **67**, 105

1,2-Benzenediol; (120-80-9), **66**, 184; **67**, 98

Benzenemethanamine, N-(methoxymethyl)-N-[(trimethylsilyl)methyl]-, **67**, 133

Benzenemethanamine, N-[(trimethylsilyl)methyl]-, **67**, 133

Benzenemethanaminium, N,N,N-trimethyl-, hydroxide; (100-85-6), **68**, 56

Benzenemethanol, α-methyl-3-nitro-, **67**, 180

Benzenemethanethiol, **65**, 215

Benzenepropanol, 2-amino-, (S)-; (3182-95-4), **68**, 77

Benzeneseleninic acid, **67**, 157

Benzenesulfenyl chloride; (931-59-9); **68**, 8

Benzenesulfonamide, **66**, 203

Benzenesulfonic acid, hydrazide, **67**, 157

Benzenesulfonoselenoic acid, Se-phenyl ester, **67**, 157

1-(BENZENESULFONYL)CYCLOPENTENE: BENZENE, (1-CYCLOPENTEN-

     1-YLSULFONYL)-; (64740-90-5), **67**, 157, 163

Benzenesulfonyl hydrazide: Benzenesulfonic acid, hydrazide; (80-17-1), **67**, 157

Benzenethiol; (108-98-5), **68**, 8

2H-1-BENZOPYRAN-3-CARBOXYLIC ACID, 5,6,7,8-TETRAHYDRO-2-OXO-,

     METHYL ESTER, **65**, 98

p-Benzoquinone; (106-51-4), **67**, 105; **68**, 109

4H-1,3-BENZOXATHIIN, HEXAHYDRO-4,4,7-TRIMETHYL-, **65**, 215

Benzoyl chloride, 4-nitro-, **67**, 86

Benzyl alcohol, α-methyl-m-nitro-, **67**, 180

Benzylamine; (100-46-9), **67**, 105, 133; **68**, 227

Benzylamine hydrochloride; (3287-99-8), **68**, 206

N-BENZYL-2-AZANORBORNENE: 2-Azabicyclo[2.2.1]hept-5-ene, 2-(phenylmethyl)-),

     68, 206

Benzyl bromide; (100-39-0), **68**, 92

2,2'-Bis(diphenylphosphino)-1,1'-binaphthyl-(η^4-1,5-cyclooctadiene)-
   rhodium(I) perchlorate, (+)- and (-): Rhodium(1+), [[1,1'-binaph-
   thalene]-2,2'diylbis[diphenylphosphine]-P,P'][(1,2,5,6-η)–1,5-cyclo-
   octadiene]-, stereoisomer perchlorate; (+)-; (82822-45-5); (-)-;
   (82889-98-3), **67**, 33

2,2'-Bis(diphenylphosphinyl)-1,1'-binaphthyl, (±)-, (S)-(-)-, and (R)-(+)- [(±)-,
   (S)-(-)-, and (R)-(+)-BINAPO]: Phosphine oxide, [1,1'-binaphthalene]-
   2,2'-diylbis[diphenyl-, (±)-, (S)-, and (R)-]; (±)-; (866632-33-9); (S)-;
   (94041-18-6); (R)-; (94041-16-4), **67**, 20

1,2-Bis(2-iodoethoxy)ethane: Ethane, 1,2-bis(2-iodoethoxy)-; (36839-55-1), **68**, 227

N^{1}N^{2}Bis(methoxycarbonyl)sulfur diimide: Sulfur diimide, dicarboxy-,
   dimethyl ester; (16762-82-6), **65**, 159

1,2-Bis(tributylstannyl)ethylene, (E)-: Stannane, vinylenebis[tributyl-, (E)-;
   Stannane, 1,2-ethenediylbis[dibutyl-, (E)-; (14275-61-7), **67**, 86

Bis(tributyltin) oxide: Distannoxane, hexabutyl-; (56-35-9), **68**, 104

1,2-Bis(trimethylsiloxy)cyclobut-1-ene: Silane, (1-cyclobuten-1,2-
   ylenedioxy)bis[trimethyl-; Silane, [1-cyclobutene-1,2-
   diylbis(oxy)]bis[trimethyl-; (17082-61-0), **65**, 17

1,4-BIS(TRIMETHYLSILYL)BUTA-1,3-DIYNE: 2,7-DISILAOCTA-3,5-DIYNE,
   2,2,7,7-TETRAMETHYL-; SILANE, 1,3-BUTADIYNE-1,4-DIYLBIS-
   [TRIMETHYL-; (4526-07-2), **65**, 52

BNP acid, R(-) : R(-) Dinaphtho[2,1-d:1'2'-f][1,3,2]dioxaphosphepin,
   4-hydroxy-4-oxide, **67**, 1
   Compound with 8α, 9–cinchonan-9-ol; (40481-36-5), **67**, 1

BNP acid, S(+) : S(+) Dinaphtho[2,1-d:1'2'-f][1,3,2]dioxaphosphepin,
   4-hydroxy-4-oxide, **67**, 1
   Compound with 9S-cinchonan-9-ol; (3974950-3), **67**, 1

Borane-dimethylsulfide; (13292-87-0), **68**, 77

Borate(1-), tetrafluoro-, hydrogen, compound with oxybis[methane], **67**, 141

Boron trifluoride etherate: Ethyl ether, compd. with boron fluoride (BF$_3$) (1:1);
Ethane, 1,1'-oxybis-, compd. with trifluoroborane (1:1); (109-63-7), **65**, 17I;
67, 52, 205; **68**, 77

4-Bromobutanenitrile: Butyronitrile, 4-bromo-; Butanenitrile, 4-bromo-;
(5332-06-9), **67**, 193

2-Bromo-2-butene (cis and trans mixture): 2-Butene, 2-bromo-; (13294-71-8), **65**, 42

1-Bromo-3-chloro-2,2-dimethoxypropane: 2-Propanone, 1-bromo-3-chloro-,
dimethyl acetal; Propane, 1-bromo-3-chloro-2,2-dimethoxy-; (22089-54-9), **65**, 2

1-Bromo-1-chloromethane, **67**, 76

6-Bromo-3,4-dimethoxybenzaldehyde: Benzaldehyde, 2-bromo-4,5-dimethoxy-;
(5392-10-9), **65**, 108

6-Bromo-3,4-dimethoxybenzaldehyde cyclohexylimine: Cyclohexanamine
N-[(2-bromo-4,5-dimethoxyphenyl)methylene]-; (73252-55-8), **65**, 108

1-BROMO-1-ETHOXYCYCLOPROPANE: CYCLOPROPANE, 1-BROMO-1-
ETHOXY-; (95631-62-2), **67**, 210

Bromoform; (75-25-2), **68**, 220

BROMOMETHANESULFONYL BROMIDE; METHANESULFONYL BROMIDE,
BROMO-; (54730-18-6), **65**, 90

2-(Bromomethyl)-2-(chloromethyl)-1,3-dioxane: 1,3-Dioxane, 2-(bromomethyl)-
2-(chloromethyl)-; (60935-30-0), **65**, 32

N-Bromomethylphthalimide: Phthalimide, N-(bromomethyl)-; 1H-Isoindole-
1,3-(2H)-dione, 2-(bromomethyl)-; (5332-26-3), **65**, 119

1-Bromo-4-nitrobenzene, **66**, 68

(Z)-β-Bromostyrene: Styrene, β-bromo-, (Z)-; (588-73-8), **68**, 130

3-Butyn-2-ol; (65337-13-5), **67**, 141

CITRONELLAL, (R)-(+)-: 6-OCTENAL, 3,7-DIMETHYL, (R)-(+)-; (2385-77-5), **67**, 33

Claisen rearrangement, of propargyl alcohols, **66**, 22

Cobalt(1+), hexacarbonyl[μ[(2,3-η:2,3-η)-1-methyl-2-propynylium]]di-, tetra-fluoroborate(1-), **67**, 141

Cobalt, octacarbonyldi-, **67**, 141

CONJUGATE ADDITION/CYCLIZATION OF A CYANOCUPRATE, **66**, 52

CONJUGATE ADDITION, ZINC HOMOENOLATES TO ENONES, **66**, 43, 50

CONJUGATED DIENES, SYNTHESIS OF, **66**, 60, 64

Copper(I) bromide (7787-70-4), **65**, 203; **66**, 2, 3, 44, 45

Copper, bromo[thiobis[methane]]-; (54678-23-8), **66**, 51

Copper chloride (CuCl); (7758-89-6), **66**, 180, 182, 184; **67**, 121

Copper(I) chloride - tetramethylethylenediamine complex, **65**, 52

Copper, compound with zinc (1:1), **67**, 98

Copper cyanide (CuCN), **66**, 53, 55

Copper iodide (CuI), **66**, 97, 100, 116, 118

Crotonaldehyde; (123-73-9), **67**, 210

Cumene hydroperoxide: Hydroperoxide, α,α-dimethylbenzyl; hydroperoxide, 1-methyl-1-phenylethyl; (80-15-9), **68**, 49

Cuprate addition, **66**, 52, 95

Cuprate, dibutyl-, lithium, reaction with acetylene, **66**, 62

Cuprates, higher order, **66**, 57

Cuprous bromide/dimethyl sulfide; (54678-23-8), **66**, 51; **68**, 198

Cuprous chloride-pyridine complex, **66**, 182

[2 + 2] CYCLOADDITION, **65**, 135; **68**, 32

Cyclobutanecarboxamide; (1503-98-6), **66**, 132, 134, 136, 141

Cyclobutanecarboxylic acid; (3721-95-7), **66**, 133, 135, 141

Cyclopentanone, 2-ethenyl-2-methyl; (88729-76-4), **66**, 94

Cyclopentene; (142-29-0), **67**, 157

4-Cyclopentene-1,3-diol, monoacetate, cis-, **67**, 114

2-Cyclopenten-1-one, 4,4-dimethyl-, **67**, 121, 205

2-CYCLOPENTEN-1-ONE, 3-METHYL-2-PENTYL-, **65**, 26

CYCLOPROPANATION, **67**, 176

>with dibromocarbene, **68**, 220

Cyclopropane, 1-bromo-1-ethoxy-, **67**, 210

CYCLOPROPANE -1,2-DICARBOXYLIC ACID, (+)-(1S,2S)-: 1,2-CYCLOPRO-

>PANEDICARBOXYLIC ACID, (S,S)-(+)-; 1,2-CYCLOPROPANEDI-

>CARBOXYLIC ACID, (1S-trans)-; (14590-54-6) **67**, 76

1,2-Cyclopropanedicarboxylic acid, (S,S)-(+)-, **67**, 76

1,2-Cyclopropanedicarboxylic acid, bis[5-methyl-2-(1-methylethyl)cyclohexyl]

>ester, **67**, 76

Cyclopropane, 1-trimethylsiloxy-1-ethoxy-, **66**, 44

Cyclopropanecarboxylic acid chloride, **66**, 176

CYCLOPROPENONE 1,3-PROPANEDIOL KETAL: 4,8-DIOXASPIRO[2.5]OCT-1-ENE;

>(60935-21-9), **65**, 32

Cyclopropylpropiolic acid ethyl ester, **66**, 177

Davis reagent, **66**, 203

Decanoic acid, ethyl ester, **67**, 125

Decanoic acid, 2-(methyldiphenylsilyl)-, ethyl ester, **67**, 125

Decanoic acid, 6-oxo-; (4144-60-9), **66**, 126

Decanoic acid, 6-oxo-, methyl ester; (61820-00-6), **66**, 120

Decarboxylation, of β-keto acid, **68**, 210

(E)-1-Decenyldiisobutylalane, **66**, 60, 61

Dibromocarbene; **68**, 220

Dibromomethane: Methane, dibromo-; (74-95-3), **65**, 81

1,2-Di-tert-butoxy-1-chloroethene, (E)-: Propane, 2,2'-[(1-chloro-1,2-
ethenediyl)bis(oxy)]bis[2-methyl-, (E); (70525-93-8), **65**, 58

1,2-Di-tert-butoxy-1,2-dichloroethane, dl-: Propane, 2,2'-[(1,2-dichloro-1,2-
ethanediyl)]bis(oxy)bis[2-methyl-, (R*,R*)-(±)-; (68470-80-4), **65**, 68

1,2-Di-tert-butoxy-1,2-dichloroethane, meso-: Propane, 2,2'-[(1,2-dichloro-
1,2-ethanediyl)bis(oxy)]bis[2-methyl-, (R*,S*)-; (68470-81-5), 65, 68

2,3-Di-tert-butoxy-1,4-dioxane, cis-: 1,4-Dioxane, 2,3-bis(1,1-
dimethylethoxy)-, cis-; (68470-78-0), **65**, 68

2,3-Di-tert-butoxy-1,4-dioxane, trans-: 1,4-Dioxane,
2,3-bis(1,1-dimethylethoxy)-, trans-; (68470-79-1), 65, 68

DI-TERT-BUTOXYETHYNE; PROPANE, 2,2'-[1,2-ETHYNEDIYLBIS(OXY)]BIS[2-
METHYL-; (66478-63-5), **65**, 68

Dibutylboron triflate: Methanesulfonic acid, trifluoro-, anhydride with dibutylborinic
acid; (60669-69-4), **68**, 83

2,6-Di-tert-butyl-4-methylpyridine: Pyridine, 2,6-bis(1,1-dimethylethyl)-4-methyl;
(38222-83-2), **68**, 116, 138

Dicarbonyl(cyclopentadienyl)diiron, [(CO)$_2$CpFe]$_2$; (12154-95-9),
66, 96, 99, 106

Dicarbonyl(cyclopentadienyl)(ethyl vinyl ether)iron tetrafluoroborate,
66, 96, 106

Dicarbonyl(cyclopentadienyl)(trans-3-methyl-2-vinylcyclohexanone)iron
tetrafluoroborate, **66**, 97

Di-μ-chlorobis(η^4-1,5-cyclooctadiene)dirhodium(I): Rhodium, di-μ-chlorobis-
[(1,2,5,6-η)-1,5-cyclooctadiene]di-; (12092-47-6), **67**, 33

Dichlorobis(triphenylphosphine)palladium(II): Palladium,
dichlorobis(triphenylphosphine)-; (13965-03-2), **68**, 130

2,3-Dichloro-1,4-dioxane, trans-: 1,4-Dioxane, 2,3-dichloro-, trans-;
(3883-43-0), **65**, 68

(I,I-Dichloroiodo)benzene, **66**, 137

Dichloroketene, **68**, 32

2,4-DICHLOROMETHOXYBENZENE: ANISOLE, 2,4-DICHLORO-; BENZENE,
2,4-DICHLORO-1-METHOXY-; (553-82-2), **67**, 222

7,7-Dichloro-1-methylbicyclo[3.2.0]heptan-6-one: Bicyclo[3.2.0]heptan-6-one,
7,7-dichloro-1-methyl-; (51284-43-6), **68**, 41

Dicobalt octacarbonyl: Cobalt, octacarbonyldi-, (Co-Co); (15226-74-1), **67**, 141

5,5-Dicyano-4-phenylcyclopent-2-enone 1,3-propanediol ketal:
6,10-Dioxaspiro[4,5]dec-3-ene-1,1-dicarbonitrile, 2-phenyl-;
(88442-12-0), **65**, 32

Dicyclopentadiene, **66**, 99

Dieckmann cyclization, **66**, 52

DIELS-ALDER REACTION:

domino, **68**, 198

immonium ion-based, **68**, 206

INVERSE ELECTRON DEMAND, **66**, 142, 147, 148

of triethyl 1,2,4-triazine-3,5,6-tricarboxylate, **66**, 150

Diels-Alder reactions, **66**, 40

1,3-Diene synthesis, via palladium-catalyzed coupling, **68**, 116, 130

Diethylamine, **66**, 145; **67**, 44, 48, 105

Diethyl aminomethylphosphonate: Phosphonic acid, (aminomethyl)-,
diethyl ester; (50917-72-1), **65**, 119

DIETHYL N-BENZYLIDENEAMINOMETHYLPHOSPHONATE: PHOSPHONIC ACID,

9,10-Dihydrofulvalene: Bi-2,4-cyclopentadien-1-yl; (21423-86-9), **68**, 198

DIHYDROJASMONE, **65**, 26

1,2-Dihydroxybenzene; (120-80-9), **66**, 180, 184

Dihydroxytartaric acid disodium salt hydrate; (866-17-1), **66**, 144, 146, 149

Diiodomethane; (75-11-6), **67**, 176

Diisobutylaluminum hydride; (1191-15-7), **66**, 60, 62, 66, 186, 188, 193

Diisopropylamine; 2-Propanamine, N-(1-methylethyl)-; (108-18-9), **65**, 98;
 66, 37, 39, 88, 194; **67**, 125; **68**, 210

DIISOPROPYL (2S,3S)-2,3-O-ISOPROPYLIDENETARTRATE: 1,3-DIOXOLANE-4,5-
 DICARBOXYLIC ACID, 2,2-DIMETHYL-, BIS(1-METHYLETHYL) ESTER,
 (4R-TRANS)-; (81327-47-1), **65**, 230

2,3-Di-O-isopropylidene-L-threitol: 1,3-Dioxolane-4,5-dimethanol, 2,2-dimethyl-,
 (4S-trans)-; (50622-09-8), **68**, 92

Diketene: 2-Oxetanone, 4-methylene-; (674-82-8), **66**, 109, 111, 114; **68**, 210

Dimenthyl (1S,2S)-cyclopropane-1,2-dicarboxylate, (-)-: 1,2-Cyclopropane
 dicarboxylic acid, bis[5-methyl-2-(1-methylethyl)cyclohexyl] ester, [1S-
 [1S*,2S*,5R*)], 2β,5α]]-; (96149-01-8), **67**, 76

DIMENTHYL SUCCINATE, (-)-: BUTANEDIOIC ACID, BIS[5-METHYL-2-
 (1-METHYLETHYL)-CYCLOHEXYL]ESTER, [1R-[1α(1R*,2S*,5R*), 2β,
 5α)]-; (34212-59-4), **67**, 76

1,1-Dimethoxyethylene: Ethene, 1,1-dimethoxy-; (922-69-0), **65**, 98

2,2-Dimethoxypropane: Propane, 2,2-dimethoxy-; (77-76-9), **68**, 92

6,7-Dimethoxy-1,2,3,4-tetrahydro-2-[(1-tert-butoxy-3-methyl)-2-butylimino-
 methyl]isoquinoline: Isoquinoline 2-[[[1-[(1,1-dimethylethoxy)methyl]-2-
 methylpropyl]imino]methyl]-1,2,3,4-tetrahydro-6,7-dimethoxy-, (S)-;
 (90482-03-4), **67**, 60

6,7-Dimethoxy-1,2,3,4-tetrahydroisoquinoline hydrochloride: Isoquinoline,

    1,2,3,4-tetrahydro-6,7-dimethoxy-, hydrochloride; (2328-12-3), **67**, 60

N,N-Dimethylacetamide: Acetamide, N,N-dimethyl-; (127-19-5), **67**, 98

Dimethyl acetylenedicarboxylate: Acetylenedicarboxylic acid, dimethyl ester;

    (762-42-5), **68**, 198

3,3-Dimethylallyl bromide; (870-63-3), **66**, 76, 78, 85

Dimethylamine; (124-40-3), **68**, 162

4-(N,N-Dimethylamino)pyridine: Pyridine, 4-(dimethylamino)-;

    4-Pyridinamine, N,N-dimethyl-; (1122-58-3), **65**, 12; **68**, 83, 210

N,N-DIMETHYL-N'-(1-tert-BUTOXY-3-METHYL-2-BUTYL)FORMAMIDINE,

    (S)-: METHANIMIDAMIDE, N'-[1-[(1,1-DIMETHYLETHOXY)METHYL]-2-

    METHYLPROPYL]-N,N-DIMETHYL-, (S)-; (90482-06-7), **67**, 52, 60

N,N-Dimethylchloromethylenammonium chloride; (3724-43-4), **66**, 121, 122, 124, 126

2,2-Dimethylcyclohexane-1,3-dione: 1,3-Cyclohexanedione, 2,2-dimethyl-;

    (562-13-0), **68**, 56

4,4-DIMETHYL-2-CYCLOPENTEN-1-ONE: 2-CYCLOPENTEN-1-ONE,

    4,4-DIMETHYL-; (22748-16-9), **67**, 121, 205

Dimethyl diazomethylphosphonate: Phosphonic acid, (diazomethyl)-, dimethyl ester;

    (27491-70-9), **65**, 119

2,2-Dimethyl-1,3-dioxane-4,6-dione, **67**, 170

3,3-DIMETHYL-1,5-DIPHENYLPENTANE-1,5-DIONE: 1,5-PENTANEDIONE,

    3,3-DIMETHYL-1,5-DIPHENYL-; (42052-44-8), **65**, 12

1,2-DIMETHYLENECYCLOHEXANE: CYCLOHEXANE, 1,2-DIMETHYLENE;

    CYCLOHEXANE, 1,2-BIS(METHYLENE)-; (2819-48-9), **65**, 90

N,N-Dimethylethanolamine: Ethanol, 2-(dimethylamino)-; (108-01-0), **68**, 162

N,N-Dimethylformamide; (68-12-2), **66**, 121, 123, 195, 197; **68**, 109

N,N-Dimethylformamide dimethyl acetal: Trimethylamine, 1,1-dimethoxy-;
Methanamine, 1,1-dimethoxy-N,N-dimethyl-; (4637-24-5), **67**, 52

Dimethyl 3,3a,3b,4,6a,7a-hexahydro-3,4,7-metheno-7H-cyclopenta[a]pentalene-
7,8-dicarboxylate: 3,4,7-Metheno-7H-cyclopenta[a]pentalene-7,8-dicarboxylic
acid, 3,3a,3b,4,6a,7a-hexahydro-, dimethyl ester; (53282-97-6), **68**, 198

Dimethyl (E)-2-hexenedioate; (70353-99-0), **66**, 52, 53, 59

N,O-Dimethylhydroxylamine hydrochloride: Methylamine, N-methoxy-, hydrochloride; Methanamine, N-methoxy-, hydrochloride; (6638-79-5), **67**, 69

N,N-Dimethylisobutyramide, **66**, 117, 118

Dimethyl (2S,3S)-2,3-0-isopropylidenetartrate: 1,3-Dioxolane-4,5-dicarboxylic
acid, 2,2-dimethyl-, dimethyl ester, (4R-trans)- or (4S-trans)-;
(37031-29-1) or (37031-30-4), **65**, 230; **68**, 92

Dimethyl malonate; (108-59-8), **66**, 76, 78, 85

Dimethyl methoxymethylenemalonate: Malonic acid, (methoxymethylene)-,
dimethyl ester; Propanedioic acid, (methoxymethylene)-, dimethyl ester;
(22398-14-7), **65**, 98

1,3-DIMETHYL-3-METHOXY-4-PHENYLAZETIDINONE: 2-AZETIDINONE,
3-METHOXY-1,3-DIMETHYL-4-PHENYL-; (82918-98-7), **65**, 140

1,3-Dimethyl-5-oxobicyclo[2.2.2]octane-2-carboxylic acid, **66**, 38

2,2-Dimethyl-4-oxopentanal: Pentanal, 2,2-dimethyl-4-oxo-; (61031-76-3),
67, 121

N,N'-Dimethylpropyleneurea (DMPU); (7226-23-5), **66**, 45, 91, 94

Dimethyl sulfate; (77-78-1), **67**, 13

Dimethyl sulfide, **66**, 211

Dimethyl sulfoxide, **66**, 15, 17

(1R)-9,9-DIMETHYLTRICYCLO[6.1.1.0^{2,6}]DECA-2,5-DIENE, **68**, 220

1,9-DIMETHYL-8-(TRIMETHYLSILYL)BICYCLO[4.3.0]NON-8-EN-2-ONE, **66**, 8

ETHYNYL p-TOLYL SULFONE: BENZENE, 1-(ETHYNYLSULFONYL)-4-
METHYL-; (13894-21-8), **67**, 149

(-)-Fenchone: 2-Norbornanone, 1,3,3-trimethyl-; Bicyclo[2.2.1]heptan-2-one, 1,3,3-
trimethyl-; (1195-79-5), **68,**14

Finkelstein reaction, **66**, 87

Flash chromatography, **66**, 135, 196

Florisil, **66**, 197

Formaldehyde, **66**, 220; **68**, 206, 234, 238

Formamide, N-[1-[(1,1-dimethylethoxy)methyl]-2-methylpropyl]-, (S)-, **67**, 52

Formamidines, N,N-dimethyl-N'-alkyl-, **67**, 52

Formic acid; (64-18-6), **68**, 109, 162

Formic acid, chloro-, ethyl ester; (541-41-3), **66**, 141; **67**, 86

Formic acid, chloro-, methyl ester, **65**, 47

Formic acid, cyano-, ethyl ester; (623-49-4), **66**, 149

Formic acid, ethyl ester, **67**, 52

Formylation, of p-tert-butylphenol, **68**, 234, 238, 243

N-Formyl-O-tert-butylvalinol, (S)-: Formamide, N-[1-[(1,1-dimethylethoxy)-
methyl]-2-methylpropyl]-, (S)-; (90482-04-5), **67**, 52

2-Furancarboxaldehyde; (98-01-1), **68**, 162

2-(5H)-Furanone; (497-23-4), **68**, 162

2-(5H)-Furanone, 3-methyl-; (2122-36-7), **68**, 162

Furfural: 2-Furaldehyde; (98-01-1), **68**, 162

Furyl phosphorodichloridate: Phosphorodichloridic acid, 2-furanyl ester;
(105262-70-2), **68**, 162

Furyl N,N,N',N'-tetramethyldiamidophosphate: Phosphorodiamidic acid, tetramethyl-,
2-furanyl ester; (105262-58-6), **68**, 162

HEXACARBONYL(PROPARGYLIUM)DICOBALT SALTS, **67**, 141

Hexachloroethane; (67-72-1), **66**, 195, 197, 202

(5Z,7E)-5,7-HEXADECADIENE, **66**, 60, 61, 63

3,3A,3B,4,6A,7A-HEXAHYDRO-3,4,7-METHENO-7H-CYCLOPENTA[A]PENTALENE-
DICARBOXYLIC ACID: 3,4,7-Metheno-7H-cyclopenta[a]pentalene-
7,8-dicarboxylic acid, 3,3a,3b,4,6a,7a-hexahydro-; (61206-25-5), **68**, 198

HEXAHYDRO-4,4,7-TRIMETHYL-4H-1,3-BENZOXATHIIN: 4H-1,3-BENZOXATHIIN,
HEXAHYDRO-4,4,7-TRIMETHYL-; (59324-06-0), **65**, 215

Hexamethylphosphoric triamide (HMPA); (680-31-9), **66**, 44, 45, 51, 88, 94

Hexanedioic acid, monomethyl ester; (627-91-8), **66**, 120

2,4-Hexenedioic acid, monomethyl ester (Z,Z)-; (61186-96-7), **66**, 184

1-Hexene, 1-iodo-, (E); (16644-98-7), **66**, 66

1-Hexene, 1-iodo-, (Z); (16538-47-9), **66**, 66

(E)-1-Hexenyl-1,3,2-benzodioxaborole: 1,3,2-Benzodioxaborole, 2-(1-hexenyl)-, (E)-;
37490-22-5), **68**, 130

(E)-1-Hexenyldiisobutylalane; (20259-40-9), **66**, 66

(E)-1-Hexenyl iodide; (16644-98-7), **66**, 63, 66

(Z)-1-Hexenyl iodide; (16538-47-9), **66**, 61, 62, 66

1-Hexyne; (693-02-7), **66**, 66; **68**, 32, 130

reaction with diisobutylaluminum hydride, **66**, 63

Higher order cuprates, **66**, 57

HOFMANN REARRANGEMENT UNDER MILDLY ACIDIC CONDITIONS, **66**, 132

Homoenolate, copper, **66**, 47

Homoenolate, titanium, **66**, 47

HOMOENOLATE, ZINC, **66**, 43

Horner-Wadsworth-Emmons reaction, **66**, 220

Hydrazine, **66**, 143; **67**, 60, 187

N-Hydroxymethylphthalimide: Phthalimide, N-(hydroxymethyl)-;

    1H-Isoindole-1,3-(2H)-dione, 2-(hydroxymethyl)-; (118-29-6), **65**, 119

(2S*,3S*)-3-HYDROXY-3-PHENYL-2-METHYLPROPANOIC ACID, **68**, 83

(1-Hydroxy-2-propenyl)trimethylsilane, **66**, 14, 15

Hydrozirconation of alkynes, **66**, 64

Hypervalent iodine compounds, **66**, 138

Imines, reduction by diisobutylaluminum hydride, **66**, 189

5H-Indene-5-one, 1,2,3,3a,4,6-hexahydro-, **67**, 163

INDOLE, 4-NITRO-, **65**, 146

INTRAMOLECULAR ACYLATION OF ALKYLSILANES, **66**, 87

INVERSE ELECTRON DEMAND DIELS-ALDER, **65**, 98

Iodine; (7553-50-2), **68**, 198

Iodine, hypervalent, **68**, 175

Iodine, phenylbis(trifluoroacetato-O); (2712-78-9), **66**, 141

Iodobenzene diacetate, **66**, 136

Iodobenzene dichloride, **66**, 137

6-Iodo-3,4-dimethoxybenzaldehyde cyclohexylimine: Cyclohexanamine,

    N-[(2-iodo-4,5-dimethoxyphenyl)methylene]-; (61599-78-8), **65**, 108

Iodomethane; (74-88-4), **68**, 56, 116, 162

Iodosobenzene diacetate: Benzene, (diacetoxyiodo)-; (3240-34-4), **68**, 175

o-Iodotoluene; (615-37-2), **66**, 67, 68, 74

1-Iodo-3-trimethylsilylpropane; (18135-48-3), **66**, 87, 88, 91, 94

Ion exchange chromatography, **66**, 212, 214, 215

Iron pentacarbonyl, **66**, 99

Isobutene; (115-11-7), **67**, 52

Isobutyl chloroformate, **66**, 135

(R)-Isobutyloxirane, **66**, 165

Isoindole-1,3-(2H)-dione, 2-(bromomethyl)-, 1H-, **65**, 119

Isoindole-1,3-(2H)-dione, 2-(hydroxymethyl)-, 1H-, **65**, 119

(S)-Isoleucine, **66**, 153

Isomenthol, (+)-: Cyclohexanol, 5-methyl-2-(1-methylethyl)-, [1S-(1α,2β,5β)];

    (23283-97-8), **65**, 81

Isomenthone, (+)-: Cyclohexanone, 5-methyl-2-(1-methylethyl)-, (2R-cis)-;

    (1196-31-2), **65**, 81

ISOPRENE; (78-79-5), **67**, 48

Isopropyl alcohol, titanium (4+) salts; (546-68-9), **65**, 230; **67**, 180; **68**, 49

Isopropylideneacetophenone: 2-Buten-1-one, 3-methyl-1-phenyl-;

    (5650-07-7), **65**, 12

Isopropylidene α-(hexahydroazepinylidene-2)malonate: 1,3-Dioxane-4,6-

    dione, 5-(hexahydro-2H-azepin-2-ylidene)-2,2-dimethyl-; (70192-54-8),

    67, 170

(R)-Isopropyloxirane, **66**, 165

Isoquinoline, 1,2,3,4-tetrahydro-6,7-dimethoxy, hydrochloride, **67**, 60

Isoquinoline, 1,2,3,4-tetrahydro-6,7-dimethoxy-1-methyl-, (S)-, **67**, 60

Jones' reagent, **68**, 175

Ketones, preparation from carboxylic acid, **66**, 119

β–Lactams, **65**, 140

Lactic acid, 2-methyl-, methyl ester; (2110-78-3), **66**, 114

Lead dioxide: Lead oxide; (1309-60-0), **65**, 166

Lead oxide, **65**, 166

Magnesium ethoxide, **66**, 175

Maleimide, N-phenyl-, **67**, 133

Malonic acid, cyclic isopropylidene ester, **67**, 170

Malonic acid, dimethyl ester; (108-59-8), **66**, 85

Malonic acid, (methoxymethylene)-, dimethyl ester, **65**, 98

Malonic ester alkylation, **66**, 75

Malononitrile, benzylidene-, **65**, 32

Manganese dioxide; (1313-13-9), **68**, 109

MELDRUM'S ACID: 2,2-DIMETHYL-1,3-DIOXANE-4,6-DIONE; MALONIC
    ACID, CYCLIC ISOPROPYLIDENE ESTER; 1,3-DIOXANE-4,6-DIONE,
    2,2-DIMETHYL-; (2033-24-1), **67**, 170

p-Mentha-6,8-dien-2-one; (6485-40-1), **66**, 13

p-MENTH-4-(8)-EN-3-ONE, (R)-(+)-, **65**, 203, 215

Menthol, (-)-: Cyclohexanol, 5-methyl-2-(1-methylethyl)-, [1R-(1α,2β,5α)]—;
    (2216-51-5), **67**, 76; **68**, 155

(-)-MENTHYL CINNAMATE: 2-Propenoic acid, 3-phenyl-, 5-methylethyl)cyclohexyl
    ester, [1R-(1α,2β,5α)]-; (16205-99-5), **68**, 155

(-)-MENTHYL NICOTINATE: 3-Pyridinecarboxylic acid, 5-methyl-2-(1-
    methylethyl)cyclohexyl ester, [1R-(1α2β,5α)]-, **68**, 155

2-(1-Mercapto-1-methylethyl)-5-methylcyclohexanol: Cyclohexanol-2-
    (1-mercapto-1-methylethyl)-5-methyl-, [1R-(1α,2α,5α)]-;
    (79563-68-1); [1R-(1α,2β,5α)] (79563-59-0); [1S-(1α,2α,5β)]-;
    (79563-67-0), **65**, 215

Mercuric chloride; (7487-94-7), **68**, 148

Mercurisulfonylation, of cyclohexa-1,3-diene, **68**, 148

Mercury, **66**, 96

Mercury(II) oxide, red; (21908-53-2), **67**, 205

METHACRYLOYL CHLORIDE: 2-PROPENOYL CHLORIDE, 2-METHYL-;

    (920-46-7), **67**, 98

Methanamine, 1,1-dimethoxy-N,N-dimethyl-, **67,** 52

Methanamine, N-methoxy-, hydrochloride, **67**, 69

Methanamine, N-(phenylmethylene)-, **65**, 140

Methane, dibromo-, **65**, 81

Methane, nitro-; (75-52-5), **68**, 8

Methane, nitro-, ion(1-), sodium; (25854-38-0), **68**, 8

METHANESULFONYL BROMIDE, BROMO-, **65**, 90

Methanesulfonyl chloride; (124-63-0), **66**, 1, 3, 186, 187, 193

Methanethiol, **66**, 188

Methanimidamide, N'-[1-[(1,1-dimethylethoxy)methyl]-2-methylpropyl]-N,N-di-

    methyl-, (S)-, **67**, 52, 60

Methanol; (67-56-1), **66**, 85, 182, 184

4-Methoxyaniline: p-Anisidine; (104-94-9), **68**, 188

6-Methoxy-7-methoxycarbonyl-1,2,3,4-tetrahydronaphthalene: 2-Naphthalene-

    carboxylic acid, 5,6,7,8-tetrahydro-3-methoxy-, methyl ester;

    (78112-34-2), **65**, 98

4-METHOXY-3-PENTEN-2-ONE: 3-PENTEN-2-ONE 4-METHOXY-;

    (2845-83-2), **67**, 202

1-(4-METHOXYPHENYL)-1,2,5,6-TETRAHYDROPYRIDINE, **68**, 188

N-(4-Methoxyphenyl)-Z-4-(trimethylsilyl)-3-butenamine, **68**, 188

1-Methoxy-3-(trimethylsiloxy)-1,3-butadiene: Silane, [(3-methoxy-1-methylene-

    2-propenyl)oxy]trimethyl-, (59414-23-2), **67**, 163

1-Methoxyvinyllithium, **68**, 25

1-(Methoxyvinyl)trimethylsilane: Silane, (1-methoxyethenyl)trimethyl-;

    (79678-01-6), **68**, 25

Methyl acrylate; (96-33-3), **66**, 54, 59

    dimerization by Pd(II), **66**, 52

Methylamine, N-benzylidene-, **65**, 140

Methylamine, N-methoxy-, hydrochloride, **67**, 69

METHYL 4-O-BENZOYL-6-BROMO-6-DEOXY-α–D-GLUCOPYRANOSIDE:

    GLUCOPYRANOSIDE, METHYL 6-BROMO-6-DEOXY, 4-BENZOATE,

    α–D-; (10368-81-7), **65**, 243

Methyl 4,6-O-benzylidene-α–D-glucopyranoside: Glucopyranoside, methyl

    4,6-O-benzylidene-α–D-; α–D-glucopyranoside, methyl 4,6-O-

    (phenylmethylene)-; (3162-96-7), **65**, 243

N-Methylbenzylidenimine: Methylamine, N-benzylidene-; Methanamine,

    N-(phenylmethylene)-; (622-29-7), **65**, 140

Methyl 4-bromo-1-butanimidate hydrochloride, **67**, 193

2-Methyl-2-butene: 2-Butene, 2-methyl-; (513-35-9), **65**, 159

(3-Methyl-2-butenyl)propanedioic acid, dimethyl ester;

    (43219-18-7), **66**, 78, 85

(3-Methyl-2-butenyl)(2-propynyl)propanedioic acid, dimethyl ester, **66**, 76

O-METHYLCAPROLACTIM: 2H-AZEPINE, 3,4,5,6-TETRAHYDRO-7-

    METHOXY-; (2525-16-8); **67**, 170

Methyl carbamate: Carbamic acid, methyl ester; (598-55-0), **65**, 159

METHYL 4-CHLORO-2-BUTYNOATE: 2-BUTYNOIC ACID, 4-CHLORO-, METHYL

    ESTER; (41658-12-2), **65**, 47

Methyl chloroformate: Formic acid, chloro-, methyl ester;

    Carbonochloridic acid, methyl ester; (79-22-1), **65**, 47

Methyl cinnamate: Cinnamic acid, methyl ester; (103-26-4), **68**, 155

Methyl (E)-crotonate; (623-43-8), **66**, 38, 39, 41, 42

2-Methylcyclohexane-1,3-dione: 1,3-Cyclohexanedione, 2-methyl-;

    (1193-55-1), **68**, 56

1-Methylcyclohexene: Cyclohexene, 1-methyl-; (591-49-1), **65**, 90

3-Methyl-2-cyclohexen-1-one; (1193-18-6), **66**, 37, 39, 42

cis-1-METHYLCYCLOPENTANE-1,2-DICARBOXYLIC ACID:

    1,2-Cyclopentanedicarboxylic acid, 1-methyl-, cis-(±)-; (70433-31-7), **68**, 41

1-Methyl-1-cyclopentene: Cyclopentene, 1-methyl-; (693-89-0), **68**, 41

Methyl N,N-dichlorocarbamate: Carbamic acid, dichloro-, methyl ester;

    (16487-46-0), **65**, 159

Methyl 1,3-dimethyl-5-oxobicyclo[2.2.2]octane-2-carboxylate, **66**, 37

Methyl 1,1'-dinaphthyl-2,2'-diyl phosphate, (R)-(-)-: Dinaphtho[2,1-d:1'2'-f]di-

    oxaphosphepin, 4-methoxy-, 4-oxide, (R)-; (86334-02-3), **67**, 13

METHYLENATION OF CARBONYL COMPOUNDS, **65**, 81

3-Methylene-4-isopropyl-1,1-cyclopentanedicarboxylic acid,

    dimethyl ester, **66**, 78

3-METHYLENE-CIS-p-MENTHANE, (+)-: (CYCLOHEXANE,

    5-METHYL-1-METHYLENE-2-(1'-METHYLETHYL)-, R,R-), **65**, 81

1-METHYL-2-ETHYNYL-endo-3,3-DIMETHYL-2-NORBORNANOL;

    (1195-79-5), **68**, 14

3-METHYL-2(5H)-FURANONE: 2(5H)-Furanone, 3-methyl-, **68**, 162

2-(3-Methylfuryl) tetramethylaminophosphate: Phosphorodiamidic acid, tetramethyl-,

    3-methyl-2-furanyl ester; (105262-59-7), **68**, 162

Methyl α–D-glucopyranoside: Glucopyranoside, methyl, α–D-;

    α–D-glucopyranoside, methyl; (97-30-3), **65**, 243

4-METHYL-3-HEPTANONE, (S)-(+)-: 3-HEPTANONE, 4-METHYL-, (S)-;

    (51532-30-0), **65**, 183

5-Methyl-2-[1-methyl-1-(phenylmethylthio)ethyl]cyclohexanone, cis- and trans-;
    65, 215

5-Methyl-2-(1-methyl-1-thioethyl)cyclohexanol, **65**, 215

N-Methylmorpholine, **66**, 133, 135

Methyl nicotinate: Nicotinic acid, methyl ester; (93-60-7), **68**, 155

2-Methyl-3-nitroaniline: o-Toluidine, 3-nitro-; Benzeneamine,
    2-methyl-3-nitro-; (603-83-8), **65**, 146

2-METHYL-4'-NITROBIPHENYL, **66**, 67, 68

(R)-METHYLOXIRANE; (15448-47-2), **66**, 160, 161, 163, 164, 172

METHYL 6-OXODECANOATE; (61820-00-6), **66**, 116, 117, 120, 123

METHYL 2-OXO-5,6,7,8-TETRAHYDRO-2H-1-BENZOPYRAN-3-CARBOXYLATE:
    2H-1-BENZOPYRAN-3-CARBOXYLIC ACID, 5,6,7,8-TETRAHYDRO-2-OXO-,
    METHYL ESTER; (85531-80-2), **65**, 98

(E)-3-Methyl-3-penten-2-one, **66**, 11

(Z)-3-Methyl-3-penten-2-one, **66**, 11

3-METHYL-2-PENTYL-2-CYCLOPENTEN-1-ONE: 2-CYCLOPENTEN-1-ONE,
    3-METHYL-2-PENTYL-; (1128-08-1), **65**, 26

2-METHYL-2-PHENYL-4-PENTENAL: 4-PENTENAL, 2-METHYL-2-PHENYL-;
    (24401-39-6), **65**, 119

N-Methylpiperidine: 1-Methylpiperidine; (626-67-5), **67**, 69

2-Methylpropane-2-thiol; (75-66-1), **66**, 108, 109, 111, 114

2-(1-METHYL-2-PROPYNYL)CYCLOHEXANONE, **67**, 141

Methyl tiglate; (6622-76-0), **66**, 88, 91, 94

Methyl p-tolyl sulfide: Sulfide, methyl p-tolyl; Benzene, 1-methyl-4-(methylthio)-;
    (623-13-2), **68**, 49

(S)-(-)-METHYL p-TOLYL SULFOXIDE: Benzene, 1-methyl-4-(methylsulfinyl)-, (S)-;
    (5056-07-5), **68**, 49

3'-NITRO-1-PHENYLETHANOL: BENZENEMETHANOL, α-METHYL-3-NITRO-;
(5400-78-2), **67**, 180

1-Nitroso-2-methoxymethylpyrrolidine, (S)-: Pyrrolidine, 2-(methoxymethyl)-
1-nitroso-, (S)-; (60096-50-6), **65**, 183

Nitrosonium tetrafluoroborate, **66**, 54

Nitroso-tert-octane: Pentane, 2,2,4-trimethyl-4-nitroso-; (31044-98-1), **65**, 166

(1R)-Nopadiene: 2-Norpinene, 6,6-dimethyl-2-vinyl-, (+); (30293-06-2), **68**, 220

(1R)-(-)-Nopol: Bicyclo[3.1.1]hept-2-ene-2-ethanol, 6,6-dimethyl-, (1R)-; (35836-73-8),
68, 220

2-Norbornanone, 1,3,3-trimethyl-; (1195-79-5), **68**, 14

1,6-Octadien-1-amine, N,N-diethyl-3,7-dimethyl, [R-(E)]-, **67**, 33

2,6-Octadien-1-amine, N,N-diethyl-3,7-dimethyl, (E)-, **67**, 33, 44

2,6-Octadien-1-amine, N,N-diethyl-3,7-dimethyl-, (Z)-, **67**, 33, 48

1,6-Octadiene, 7-methyl-3-methylene-, **67**, 44

6-Octenal, 3,7-dimethyl-, (R)-(+)-, **67**, 33

tert-Octylamine: 2-Pentanamine, 2,4,4-trimethyl-; (107-45-9), **65**, 166

N-tert-Octyl-O-tert-butylhydroxylamine: 2-Pentanamine, N-(1,1-dimethylethoxy)-
2,4,4-trimethyl-; (68295-32-9), **65**, 166

1-Octyn-3-ol; (818-72-4), **66**, 22, 23, 28

Organoaluminum compounds, reaction with imino carbocations, **66**, 189

Orthoester Claisen rearrangement, **66**, 22

Orthoformic acid, triethyl ester, **65**, 146

6-Oxabicyclo[3.1.0]hex-2-ene, **67**, 114

Oxalic acid, diethyl ester, **65**, 146

Oxalyl chloride; (79-37-8), **66**, 15, 17, 21, 89, 94, 117, 121, 123

1,3-OXATHIANE, **65**, 215

PALLADIUM-CATALYZED syn-ADDITION OF CARBOXYLIC ACIDS,
67, 114

Palladium(II) chloride; (7647-10-1), 67, 121; 68, 130

Palladium, chloro(phenylmethyl)bis(triphenylphosphine)-, 67, 86

Palladium-on-carbon, 68, 227

Palladium sponge, 66, 54

Palladium, tetrakis(acetonitrile)-, tetrafluoroborate, 66, 52

Palladium, tetrakis(triphenylphosphine)-; (14221-01-3),
66, 61, 62, 66, 68, 69, 74; 67, 86, 98, 105, 114; 68, 116

Paraformaldehyde: Poly(oxymethylene); (9002-81-7), 65, 215;
66, 220, 221; 68, 243

1,4-Pentalenediol, octahydro-, (1α,3aα,4α,6aα)-; (17572-86-0), 68, 175

1,2,3,4,5-PENTAMETHYLCYCLOPENTADIENE: 1,3-CYCLOPENTADIENE,
1,2,3,4,5-PENTAMETHYL-; (4045-44-7), 65, 42

Pentanal, 2,2-dimethyl-4-oxo-, 67, 121

2-Pentanamine, N-(1,1-dimethylethoxy)-N-(1,1-dimethylethyl)-
2,4,4-trimethyl-, 65, 166

2-Pentanamine, N-(1,1-dimethylethoxy)-2,4,4-trimethyl-, 65, 166

2-PENTANAMINE, N-(1,1-DIMETHYLETHYL)-2,4,4-TRIMETHYL-, 65, 166

2-Pentanamine, 2,4,4-trimethyl-, 65, 166

1,5-PENTANEDIONE, 3,3-DIMETHYL-1,5-DIPHENYL-, 65, 12

Pentane, 2,2,4-trimethyl-4-nitroso-, 65, 166

3-Pentanone SAMP-hydrazone: 1-Pyrrolidinamine, N-(1-ethylpropylidene)-2-
(methoxymethyl)-, (S)-; (59983-36-7), 65, 183

4-PENTENAL, 2-METHYL-2-PHENYL-, 65, 119

3-Penten-2-one, 4-methoxy-, 67, 202

Perchloric acid, silver(1+) salt, monohydrate, 67, 33

Performic acid, oxidation of furfural, **68**, 162

PERRILLYL ALCOHOL, (S)-(-)-: 1-CYCLOHEXENE-1-METHANOL, 4-
(1-METHYLETHENYL)-; (536-59-4), **67**, 176

L-Phenylalanine; Phenylalanine, (S)-; (63-91-2), **68**, 77

(S)-Phenylalanol; (3182-95-4), **68**, 77

Phenylbenzeneselenosulfonate: Benzenesulfonoselenoic acid, Se-phenyl
ester; (60805-71-2), **67**, 157

N-Phenylhydroxylamine: Hydroxylamine, N-phenyl-; Benzeneamine,
N-hydroxy-; (100-65-2), **67**, 187

N-Phenylmaleimide: Maleimide, N-phenyl-; 1H-Pyrrole-2,5-dione, 1-phenyl-;
(941-69-5), **67**, 133

8-PHENYLMENTHOL, (-)-: CYCLOHEXANOL, 5-METHYL-2-(1-METHYL-
1-PHENYLETHYL)-, [1R-(1α,2β,5α)]-; (65253-04-5), **65**, 203

(S)-4-(PHENYLMETHYL)-2-OXAZOLIDINONE: 2-Oxazolidinone, 4-(phenylmethyl)-,
(S); (90719-32-7), **68**, 77, 83

(1Z,3E)-1-PHENYL-1,3-OCTADIENE: Benzene, 1,3-octadienyl-, (Z,E)-; (39491-66-2),
68, 130

3-PHENYL-4-PENTENAL, **66**, 29, 31, 36

2-Phenyl-N-(phenylmethylene)-1-propen-1-amine: 1-Propen-1-amine,
2-Phenyl-N-(phenylmethylene)-; (64244-34-4), **65**, 119

(E)-3-[(E)-3-Phenyl-2-propenoxy]acrylic acid; (88083-18-5), **66**, 30, 31, 36

Phenylseleninic acid: Benzeneseleninic acid; (6996-92-5), **67**, 157

Phenylsulfenyl chloride: Benzenesulfenyl chloride; (931-59-9), **68**, 8

trans-3-(Phenylsulfonyl)-4-(choromercuri)cyclohexene: Mercury,
chloro[2-(phenylsulfonyl)-3-cyclohexen-1-yl]-, trans-; (102815-53-2), **68**, 148

2-(PHENYLSULFONYL)-1,3-CYCLOHEXADIENE: Benzene, (1,5-cyclohexadien-
1-ylsulfonyl)-; (102860-22-0), **68**, 148

Phosphorus tribromide; (7789-60-8), **67**, 210

Phthalimide, N-(bromoethyl)-, **65**, 119

Phthalimide, N-(hydroxymethyl)-, **65**, 119

PINENE, (-)-α-: 2-PINENE, (1S,5S)-(-)-; BICYCLO[3.1.1]HEPT-2-ENE,

2,6,6-TRIMETHYL-, (1S)-; (7785-26-4), **65**, 224

PINENE, (-)-β-: BICYCLO[3.1.1]HEPTANE, 6,6-DIMETHYL-2-METHYLENE-, (1S)-;

(18172-67-3), **65**, 224

Piperidine, 1-methyl, **67**, 69

Piperidine, 2,2,6,6-tetramethyl-, **67**, 76

Poly(oxy-1,2-ethanediyl), α-[4-(1,1,3,3-tetramethylbutyl)phenyl]-ω-hydroxy-;

(9002-93-1), **68**, 56

Poly(oxymethylene), **65**, 215; **68**, 188

Potassium 3-aminopropylamide (KAPA), **65**, 224

Potassium tert-butoxide, **66**, 127, 128, 195; **67**, 125

Potassium hydride; (7693-26-7), **65**, 224

Potassium hydroxide, **66**, 89

Prenyl bromide, **66**, 76

Proline, D-; (344-25-2), **65**, 173

Proline, L-; (147-85-3), **65**, 173

Propanal; (123-38-6), **68**, 64

2-Propanamine, N-ethyl-N-(1-methylethyl)-; (7087-68-5), **68**, 162

2-Propanamine, N-(1-methylethyl)-, **65**, 98

Propane, 1-bromo-3-chloro-2,2-dimethoxy-, **65**, 32

Propane, 2,2'-[(1-chloro-1,2-ethenediyl)bis(oxy)]bis[2-methyl-, (E)-, **65**, 68

Propane, 2,2'-[(1,2-dichloro-1,2-ethanediyl)]bis(oxy)bis[2-methyl-,

(R*,R*)-(±)-, **65**, 68

Propane, 2,2'-[(1,2-dichloro-1,2-ethanediyl)bis(oxy)]bis[2-methyl-,

(R*,S*)-, **65**, 68

PROPANE, 2,2'-[1,2-ETHYNEDIYLBIS(OXY)]BIS[2-METHYL-, **65**, 68

1,3-Propanediamine; (109-76-2), **65**, 224; **66**, 131

Propanedinitrile, (phenylmethylene)-, **65**, 32

Propanedioic acid, (cyclopropylcarbonyl)-, diethyl ester;

(7394-16-3), **66**, 179

Propanedioic acid, diethyl ester; (105-53-3), **66**, 179

Propanedioic acid, (methoxymethylene)-, dimethyl ester, **65**, 98

Propanedioic acid, (3-methyl-2-butenyl)-, dimethyl ester;

(43219-18-7), **66**, 85

1,3-Propanediol; (504-63-2), **65**, 32

1,3-Propanedithiol; (109-80-8), **65**, 150

2-Propanethiol, 2-methyl-; (75-66-1), **66**, 114

Propanoic acid, 2-chloro-, (S)-; (29617-66-1), **66**, 159, 172

Propanoic acid, 2-hydroxy-2-methyl-, methyl ester; (2110-78-3), **66**, 114

1-Propanol, 2-amino-3-phenyl-, L- or (S)-(-)-; (3182-95-4), **68**, 77

1-Propanol, 2-chloro-, (S)-(+)-; (19210-21-0), **66**, 172

2-Propanol, titanium (4+) salt, **65**, 230; **67**, 180; **68**, 49

2-Propanone, 1-bromo-3-chloro-, dimethyl acetal, **65**, 32

Propanoyl chloride; (79-03-8), **68**, 83

Propargyl alcohol: 2-Propyn-1-ol; (107-19-7), **67**, 193

Propargyl bromide; (160-96-7), **66**, 77, 79, 86

PROPARGYL CHLORIDE: PROPYNE, 3-CHLORO-; 1-PROPYNE, 3-CHLORO-;

(624-65-7), **65**, 47

1-Propen-1-amine, 1-chloro-N,N,2-trimethyl-; (26189-59-3), **66**, 120

1-Propen-1-amine, 2-phenyl-N-(phenylmethylene)-, **65**, 119

Silanamine, N,N-diethyl-1,1,1-trimethyl-; (996-50-9), **68**, 83

Silane, acetyltrimethyl-; (13411-48-8), **68**, 25

SILANE, 1,3-BUTADIYNE-1,4-DIYLBIS[TRIMETHYL-, **65**, 52

Silane, chloromethyldiphenyl-, **67**, 125

Silane, chloromethyltrimethyl-, **67**, 133; **68**, 1

Silane, (3-chloropropyl)trimethyl-; (2344-83-4), **66**, 94

Silane, chlorotrimethyl-, **65**, 1, 6, 61; **66**, 6, 14, 16, 21, 44, 47; **68**, 25

Silane, [1-cyclobutene-1,2-diylbis(oxy)]bis[trimethyl-, **65**, 17

Silane, (1-cyclobuten-1,2-ylenedioxy)bis[trimethyl-, **65**, 17

Silane, (1-cyclohexen-1-yloxy)trimethyl-, **67**, 141

SILANE, (DIAZOMETHYL)TRIMETHYL-; (18107-18-1), **68**, 1

Silane, [(1-ethoxycyclopropyl)oxy]trimethyl-; (27374-25-0), **66**, 51; **67**, 210

Silane, ethynyltrimethyl-, **65**, 52, 61

Silane, (3-iodopropyl)trimethyl-; (18135-48-3), **65**, 94

SILANE, (ISOPROPENYLOXY)TRIMETHYL-, **65,** 1

Silane, [(3-methoxy-1-methylene-2-propenyl)trimethyl-, **67**, 163

Silane, trichloro-, **67**, 20

SILANE, TRIMETHYL[(1-METHYLETHENYL)OXY]-, **65**, 1

Silane, trimethyl[[(4-methylphenyl)sulfonyl]ethynyl]-, **67**, 149

Silane, trimethyl (1-methyl-1,2-propadienyl)-; (74542-82-8), **66**, 7, 13

SILANE, TRIMETHYL(1-OXO-2-PROPENYL)-; (51023-60-0), **66**, 14, 21

Silane, trimethyl[(1-phenylethyl)oxy]-, **65**, 6, 12

Silane, trimethyl[(1-phenylvinyl)oxy]-, **65**, 6, 12

Silver nitrate, **66**, 111

    reaction with 1-trimethylsilyl-1-butene, **66**, 4

Silver(I) oxide; (20667-12-3), **66**, 111, 115

Silver perchlorate: Perchloric acid, silver(1+) salt, monohydrate;

    (14202-05-8), **67**, 33

Silver(I) trifluoroacetate; (2966-50-9), **66**, 115

    preparation, **66**, 111

Silylamine, N,N-diethyl-1,1,1-trimethyl-; (996-50-9), **68**, 83

Silylation, of 1-alkyne, **68**, 182

Skattebøl rearrangement, **68**, 220

Sodium; (7440-23-5), **66**, 76, 85, 96

Sodium amalgam, **66**, 96

Sodium benzenesulfinate: Benzenesulfinic acid, sodium salt; (873-55-2), **68**, 148

Sodium bis(2-methoxyethoxy)aluminum hydride: Aluminate (1-), dihydrobis-

    (2-methoxyethanalato)-, sodium; (22722-98-1), **67**, 13

Sodium bisulfite; (7631-90-5), **68**, 162

Sodium cyclopentadienide, **68**, 198

Sodium dicarbonyl(cyclopentadienyl)ferrate; (12152-20-4), **66**, 96, 107

Sodium ethoxide, **67**, 170

Sodium fluoride, aqueous, work-up for organoaluminum reaction, **66**, 188

Sodium hydride; (7646-69-7), **66**, 30, 32, 76, 79, 85, 109, 111, 114; **68**, 92, 198

Sodium iodide, **66**, 87; **68**, 227

Sodium methoxide, **66**, 76

Sodium naphthalenide, **65**, 166

Sodium nitrite, **66**, 151

Sodium nitromethylate: Methane, nitro-, ion(1-), sodium; (25854-38-0), **68**, 8

Sodium periodate: Periodic acid, sodium salt; (7790-28-5), **68**, 41

Sodium tungstate dihydrate: Tungstic acid, disodium salt, dihydrate;

    (10213-10-2), **65**, 166

Sonication, **67**, 133

    for reaction of 1,3-diaminopropane with alkali metals, **66**, 130

SPIRO[4.5]DECAN-1,4-DIONE; (39984-92-4), **65**, 17

Stannane, allyltributyl-; (24850-33-7), **68**, 104

Stannane, 1,2-ethenediylbis[dibutyl-, (E)-, **67**, 86

Stannane, ethenyltrimethyl; (754-06-3), **68**, 116

Stannane, tetrachloro-, **65**, 17

Stannane, tributyl-, **65**, 236

Stannane, tributylchloro-, **67**, 86

Stannane, tributylethynyl, **67**, 86

Stannane, tributyl-2-propenyl-; (24850-33-7), **68**, 104

Stannane, vinylenebis[tributyl-, (E)-, **67**, 86

STETTER REACTION, **65**, 26

Succinic anhydride, **67**, 76

SUCCINIMIDE, N-BROMO-, **65**, 243

Sucrose; α-D-Glucopyranoside, β-D-fructofuranosyl; (57-50-1), **68**, 56

Sulfenylation, with phenylsulfenyl chloride, **68**, 8

Sulfide, methyl p-tolyl; (623-13-2), **68**, 49

Sulfone, ethynyl p-tolyl, **67**, 149

Sulfonylation, of 1,3-diene, **68**, 148

2-Sulfonyloxaziridines, preparation, **66**, 207, 208

Sulfosalicylic acid spray for tlc plates, **66**, 216

Sulfoxide, enantioselective synthesis, from sulfide, **68**, 49

Sulfur chloride, **65**, 159

Sulfur dichloride: Sulfur chloride; (10545-99-0), **66**, 159

Sulfur diimide, dicarboxy-, dimethyl ester, **65**, 159

Sulfuryl chloride; (7791-25-5), **68**, 8

Sulfuryl chloride isocyanate, **65**, 135

Swern oxidation, **66**, 15, 18

2,2,6,6-Tetramethylpiperidine: Piperidine, 2,2,6,6-tetramethyl-; (768-66-1), **67**, 76

Tetramethyltin, **66**, 65

2,5,7,10-Tetraoxabicyclo[4.4.0]decane, cis-: p-Dioxino[2,3,-b]-p-dioxin, hexahydro-; [1,4]-Dioxino[2,3-b]-1,4-dioxin, hexahydro; (4362-05-4), **65**, 68

1,4,8,11-TETRATHIACYCLOTETRADECANE; (24194-61-4), **65**, 150

Tetravinyltin, **66**, 53, 55

Tetronic acids, **66**, 111, 112

Thiophenol: Benzenethiol; (108-98-5), **68**, 8

Thiourea: Urea, thio-; (62-56-6), **65**, 150

Tin chloride, **65**, 17

Tin tetrachloride: Tin chloride; Stannane, tetrachloro-; (7646-78-8), **65**, 17

Titanium(IV) isopropoxide: Isopropyl alcohol, titanium(4+) salt; 2-Propanol, titanium(4+) salt; (546-68-9), **65**, 230; **67**, 180; **68**, 49

Titanium, triisopropoxymethyl-, **67**, 180

Tin, tetramethyl-, **66**, 55

Tin, tetravinyl-, **65**, 53, 55

Titanium chloride, **65**, 81, 6

Titanium tetrachloride: Titanium chloride; (7550-45-0), **65**, 81, 6; **66**, 8, 9

Toluene, o-iodo-; (615-37-2), **66**, 74

p-Toluenenesulfonic acid, monohydrate; (6192-52-5), **68**; 92, 188

p-Toluenesulfonyl chloride; Benzenesulfonyl chloride, 4-methyl-; (98-59-9), **68**, 188

α–Toluenethiol, **65**, 215

o-Toluidine, 3-nitro-, **65**, 146

o-Tolyllithium, preparation, **66**, 69

3-Trimethylsilyl-2-propyn-1-ol, **66**, 1, 3

3-Trimethylsilyl-2-propyn-1-yl methanesulfonate, **66**, 2

Trimethylsilyl trifluoromethanesulfonate: Methanesulfonic acid, trifluoro-, trimethylsilyl

ester; (27607-77-8), **68**, 64

Trimethyltin chloride: Stannane, chlorotrimethyl-; (1066-45-1), **68**, 116

Trimethylvinyltin: Stannane, trimethylvinyl-; (754-06-3), **68**, 116

Triphenylphosphine: Phosphine, triphenyl-; (603-35-0), **68**, 130, 138

Tripropylaluminum; (102-67-0), **66**, 186, 188, 193

Tris(tetrabutylammonium) hydrogen pyrophosphate trihydrate;

(76947-02-9), **66**, 212, 219

Trithiane, sym-: S-Trithiane; 1,3,5-Trithiane; (291-21-4), **65**, 90

Triton B: Ammonium, benzyltrimethyl-, hydroxide; Benzenemethanaminium,

N,N,N-trimethyl-, hydroxide; (100-85-6), **68**, 56

Triton X-100: Glycols, polyethylene, mono[p-(1,1,3,3,-tetramethylbutyl)-phenyl]ether;

Poly(oxy-1,2-ethanediyl), α-[4-(1,1,3,3-tetramethylbutyl)-phenyl]-ω-hydroxy-;

(9002-93-1), **68**, 56

Tungstic acid, disodium salt, dihydrate, **65**, 166

ULLMANN REACTION, **65**, 108

2,5-Undecanedione; (7018-92-0), **65**, 26

2-Undecene, 2-methyl-, **67**, 125

Urea, thio-, **65**, 150

(S)-Valine, **66**, 153

Valinol: 1-Butanol, 2-amino-3-methyl-, (S)-; (2026-48-4), **67**, 52

Vilsmeier-Haack reagent, **66**, 121

Vibro-mixer, **65**, 52